"[*The Numbers Game*'s] examples travel well, as do the authors' lucid, unruffled style and their wholesome commitment to public enlightenment. . . . instructive and amusing."   —*The New York Times*

"A chatty, brief, brightly informative and quite possibly essential book . . . a genteel carpet-bombing of alternatively hysterical and gullible journalists who misread numbers. . . . Readers from left and right will find many an 'aha' moment to savor."   —*New York Post*

"*The Numbers Game* is a lot of fun to read. Blastland and Dilnot take interesting examples of questionable figures from both the U.S. and Britain, and use them to develop practical rules for critical thinking about numbers in the news."
    —Joel Best, author of *Damned Lies and Statistics* and *Stat-Spotting*

Praise for the UK edition
(published as *The Tiger That Isn't*)

"This very elegant book constantly sparks 'Aha!' moments as it interrogates the way numbers are handled by politics and the media."
    —Steven Poole, *The Guardian*

"Personal and practical . . . Might even cause a social revolution."
    —*The Independent*

"This delightful book should be compulsory reading for everyone responsible for presenting data and for everyone who consumes it."
    —*The Sunday Telegraph*

"Clear-eyed and concise."   —*The Times* (London)

"A very funny book . . . this is one of those math books that claims to be self-help, and on the evidence presented here, we are in dire need of it."   —*The Daily Telegraph*

Michael Blastland is a writer, broadcaster, and the creator of *More or Less,* the BBC Radio 4 show.

Andrew Dilnot, the former host of the show, is the principal of St. Hugh's College, Oxford, and was the director of England's Institute of Fiscal Studies.

THE NUMBERS GAME

THE COMMONSENSE GUIDE TO UNDERSTANDING NUMBERS IN THE NEWS, IN POLITICS, AND IN LIFE

Michael Blastland
and
Andrew Dilnot

GOTHAM BOOKS

GOTHAM BOOKS
Published by Penguin Group (USA) Inc.
375 Hudson Street, New York, New York 10014, U.S.A.

Penguin Group (Canada), 90 Eglinton Avenue East, Suite 700, Toronto, Ontario M4P 2Y3, Canada
(a division of Pearson Penguin Canada Inc.); Penguin Books Ltd, 80 Strand, London WC2R 0RL,
England; Penguin Ireland, 25 St Stephen's Green, Dublin 2, Ireland (a division of Penguin Books
Ltd); Penguin Group (Australia), 250 Camberwell Road, Camberwell, Victoria 3124, Australia
(a division of Pearson Australia Group Pty Ltd); Penguin Books India Pvt Ltd, 11 Community
Centre, Panchsheel Park, New Delhi–110 017, India; Penguin Group (NZ), 67 Apollo Drive,
Rosedale, North Shore 0632, New Zealand (a division of Pearson New Zealand Ltd); Penguin
Books (South Africa) (Pty) Ltd, 24 Sturdee Avenue, Rosebank, Johannesburg 2196, South Africa

Penguin Books Ltd, Registered Offices: 80 Strand, London WC2R 0RL, England

Published by Gotham Books, a member of Penguin Group (USA) Inc.

Previously published as a Gotham Books hardcover edition

First trade paperback printing, January 2010

1   3   5   7   9   10   8   6   4   2

Gotham Books and the skyscraper logo are trademarks of Penguin Group (USA) Inc.

The Library of Congress has catalogued the hardcover edition of this book as follows:
Blastland, Michael.
The numbers game: the commonsense guide to understanding numbers in the news,
in politics, and in life/by Michael Blastland, Andrew Dilnot.
p.   cm.
Includes index.
ISBN 978-1-592-40423-0 (hardcover)   ISBN 978-1-592-40485-8 (paperback)
1. Number concept.   2. Mathematics.   3. Statistics.   I. Dilnot, A. W.   II. Title.
QA141.15.B535 2009
510—dc22   2008030130

Printed in the United States of America
Set in Aldine   Designed by Ellen Cipriano

*To Catherine, Katey, Rosie,
Cait, Julia, Joe, and Kitty*

CONTENTS

INTRODUCTION

This book began over a pizza as an idea for a BBC radio program. Few took it seriously: "Numbers? On the radio?" In six short years the program *More or Less* became a fixture in the schedules, the skepticism wilted, and our extravagant ambition—of changing the culture of numbers in public argument—blinked into sunlight.

Listeners told of the subversive thrill of having the mental ammunition to shoot down official claims and dodgy data—regardless of the politics. They relished clarity on facts they'd not been given straight before, told in surprising, accessible ways that made them wonder, not always politely, why they'd had to wait so long for what seemed so straightforward. The program chased down bad data and sought out good to answer pressing questions about economic and social life, it poked fun at politicians, media, and others who were content to spout numerical gibberish, it sifted research and delved into surveys and samples to find the true measure of trends, attitudes, and behavior, it sought to put risk into human proportion, and to popularize simple principles and tricks for seeing through numbers. Wherever they appeared—and they seemed to appear everywhere— we insisted they speak clearly, exposing their limitations, acknowledging their uncertainty, but also applauding their insights. In doing so, we came across an apparently endless stream of stories, some comic, some tragic, some scandalous.

Neither of us is a professional mathematician or statistician. One is a Cambridge English graduate who began asking dumb questions about numbers in the news only to find that too many of the answers were even dumber, the other an economist who is now principal of a college at Oxford University, and came to public notice as the fiercely independent head of an economic research institute. One thinks the other tall enough to be a mutant giant, while the second thinks the first should get a proper job as a jockey, which makes us middling, on average, and just goes to show the trouble with averages. What we share is the same incredulity at the way a whole language seems to be debased.

The radio program acquired a growing, often devoted, always opinionated audience of nearly a million, a Web site, imitators in the press, the financial backing of the Open University, and the interest of publishers. It became part of BBC training for new and established journalists, the basis of lectures, articles in the press, and journals and, though we hope this is not the last of its manifestations, the pizza turned into a book, and sold out twice in the UK in two months.

This edition, extensively revised for American readers, has the same aims, above all to prove that what we can do, they can do, and no doubt surpass. That's just as well; numbers nowadays saturate the news, politics, life, in the United States as in the UK. For good or ill, they are today's preeminent public language—and those who speak it rule. Quick and cool, numbers seem to have conquered fact.

But they are also hated, often for the same reasons. They can bamboozle not enlighten, terrorize not guide, and all too easily end up abused and distrusted. Potent but shifty, the role of numbers is frighteningly ambiguous. How can we see through them? Our answer is unconventional:

First, relax . . .

We know more than we think we do. We have been beautifully conditioned to make sense of numbers, believe it or not, by our own experience. This is the radical premise of this book—that readers have no need to throw up their hands in fear or contempt, if only they see how much they know already.

Numbers can make sense of a world otherwise too vast and intricate to get into proportion. They have their limitations, no doubt, but are sometimes, for some tasks, unbeatable. That is, if used properly. So although there is here a rich store of mischief and scandal, it is not to discredit numbers themselves. There are lies and damned lies in statistics, for sure, but scorning numbers is no remedy. For that is to give up the game on every political, economic, or social argument you follow, every cause you love or hate.

Our aim is rather to bring numbers back to earth, not only by uncovering the tricks of the trade—the multiple counting, suspicious graphs, sneaky start dates, and funny scales—there have been exposés of that kind of duplicity before, though there are gems in the stories that follow; nor by relying on arcane statistical techniques, brilliant though those often are. Instead, wherever possible, we offer images from life—self, experience, prompts to the imagination—to show how to cut through to what matters. It is all there—all of us possess most of it already—this basic mastery of the principles that govern the way numbers work. It can be shared, we think, even by those who once found math a cobwebbed mystery.

But simple does not mean trivial; simple numbers help to answer imperative questions. Do we know what people earn and owe, who is rich and who poor? Is that government spending promise worth a dime? Who lives and who dies by performance measures for health care? Are educational ranking charts honest? Do speed cameras really save lives? What about that survey of teenage offending, the 1 in 4 who do this, the 6 percent increase in risk for women who do that, Iraqi war dead, HIV/AIDS cases, how the United States compares with other countries, the decline of fish stocks or other wildlife, the threat of cancer, health budgets, third world debt, recycling rates, predictions of global warming? Hardly a subject is broached these days without measurements, quantifications, forecasts, rankings, statistics, targets, numbers of every variety; they are ubiquitous, and often disputed. If we are the least bit serious about any of these issues, we should attempt to get the numbers straight.

This means taking on lofty critics. Too many find it is easier to distrust numbers wholesale, affecting disdain, than get to grips with them. When a well-known writer explained to us that he had heard quite enough numbers, thank you—he didn't understand them and didn't see why he should—his objection seemed to us to mask fear. Jealous of his prejudices or the few scraps of numerical litter he already possessed, he turned up his nose at evidence in case it proved inconvenient. Everyone pays for this attitude in bad policy, bad government, gobbledygook news, and it ends in lost chances and screwed-up lives.

Another dragon better slain is the attitude that, if numbers cannot deliver the whole truth straight off, they are all just opinion. That damns them with unreasonable expectation. One of the few things we say with certainty is that some of the numbers in this book are wrong. Those who expect certainty might as well leave real life behind. Everyone is making their way precariously through the world of numbers, no single number offers instant enlightenment, life is not like that, and numbers won't be either.

Still others blame statistical bean counters for a kind of crass reductionism, and think they, with subtlety and sensitivity, know better. Sometimes there is something in this, but just as often it is the other way around. Most statisticians know the limits of any human attempt to capture life in data—they have tried, after all. Statistics is far from the dry collection of facts; it is the science of making what subtle sense of the facts we can. No science could be more necessary, and those who do it are sometimes detectives of quiet ingenuity. It is others, snatching at numbers, brash or overconfident, who are more naively out of touch.

So we should shun the extremes of cynicism or fear, on the one hand, and number idolatry on the other, and get on with doing what we can. And we can do a great deal.

Most of what is here is already used and understood in some part of their lives by almost everyone; we all apply the principles, we already understand the ideas. Everyone recognizes, for example, the folly of mistaking one big wave for a rising tide and, since we can do that, perhaps to our surprise, we can unravel arguments about whether speed cameras really save lives or cut accidents. In life, we would see—of course we would see—the way that falling rice scatters and, because we can see it, we can also make simple sense of the numbers behind cancer clusters. We know the vibrancy of the colors of the rainbow and we know what we would lack if we combined them to form a bland white band in the sky. Knowing this can, as we will see, show us what an average—average income, for example— can conceal and what it can illuminate. Many know from ready experience what it costs to buy child care, and so they can know whether government spending on child care is big or small. We are, each one of us, the obvious and ideal measure of the policies aimed at us. These things we know. And each can be a model for the way numbers work. All we seek to do is reconnect what anyone can know with what now seems mysterious, reconnect numbers with images and experience from life, such that, if we have done our job, what once was baffling or intimidating will become transparent.

What follows will not be found in a textbook: even the choice and arrangement of subjects would look odd to an expert, let alone the way they are presented. Good. This is a book from the point of view of the consumer of numbers. It is short and to the point. Each chapter starts with what we see as the nub of the matter: a principle, or a vivid image. Wipe the mental slate clean of anxiety or fuzziness and inscribe instead these ideas, keep each motif in mind while reading, see how they work in practice from the stories we tell. In this way we hope to light the path toward clarity and confidence.

The alignment of power and abuse is not unique to numbers, but it is just possible that it could be uniquely challenged, and the powerless become powerful. Here's how.

COUNTING

1

USE STRAWBERRY JAM

Counting is easy when you don't have to count anything with it: 1, 2, 3 . . . it could go on forever, precise, regular, and blissfully abstract. This is how children learn to count in the kindergarten, at teacher's knee.

But for grown-ups hopeful of putting counting to practical use, it has to lose that innocence. The difference is between counting, which is easy, and counting *something*, which is anything but. Some find this confusing, and struggle to put the childhood ideal behind them.

But life is messier than a number. The first is a shifting melee, the second a box. That is why, when counting *something*, we have to squash it into a shape that fits the numbers. (In an ideal world, the process would be the other way around and numbers would describe life, not bully it into conformity.) Worst of all, the fact that all this is a struggle is soon forgotten. To avoid this mistake, and master counting in real life, renounce the memory of the classroom and follow a better guide: strawberry jam.

***

How many centenarians are there in the United States? You don't know? Neither do we. But someone does, don't they? They just go out and count 'em, surely?

"How old are you?"

"One hundred and one, if I'm a day."

"Thank you." And they tick the box.

We often talk of social statistics, especially those that seem as straightforward as age, as if a bureaucrat were poised with a clipboard, peering through every window, counting; or, better still, had some machine to do it for them. The unsurprising truth is that, for many of the statistics we take for granted, there is no such bureaucrat, no machine, no easy count, we do not all clock in, or out, in order to be recorded, there is no roll call for each of our daily activities, no kindergarten 1, 2, 3.

What there is out there, more often than not, is thick strawberry jam, through which someone with a bad back on a tight schedule has to wade—and then try to tell us how many strawberries are in it.

The following table shows that when the U.S. Census Bureau distributed forms asking people how old they were, the answer came back that there were about 106,000 centenarians in the United States in 1970.

Job done? The Bureau believes there probably were, in truth, not 106,000, but a little less than 5,000. That is, the count was roughly twenty-two times the estimated reality. Look into the detail of the 1990 Census returns, to take another example, and it turns out that more than 1,500 people said they were aged 110 or more. What's more, among whites of this declared age, nearly 70 percent said this was without mobility or personal care limitations, 30 percent said

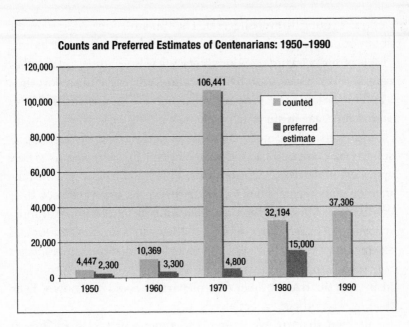

Counts and Preferred Estimates of Centenarians: 1950–1990

Source: U.S. Census Bureau

they lived alone, and about 40 percent said they were still married. There were cases where entire families were coded (by circling a date) as born in the 1800s.

The table also shows that between 1960 and 1980, the counted number of centenarians first soared to more than ten times the number of a decade earlier, then fell by more than two thirds. Proof of a sudden boom and bust in living longer? Unlikely. Rather it shows persistent overcounting, which is believed to continue to some extent to this day (the true 1990 number is estimated to be about 29,000, though some statistical commentators think it more likely to be about 22,000).

The raw data shows, in short, a mess, but a mess that reflects life.

So this is not meant as a criticism of the Bureau. Rather, it is a reflection of all of us, more confusing and indistinct than the ordinary notion of counting ever suggests. The Census Bureau describes the count of people at older ages as "hounded" by problems, including: "Lack of birth records . . . low literacy levels . . . functional and/or cognitive disability . . . individuals who deliberately misreport their age for a variety of reasons, as well as those who mistakenly report an incorrect age in surveys."

Bureau staff did not, of course, stop the effort of counting as soon as the computer had totaled the figures for centenarians as they appeared on returned Census forms. This was the point at which the investigation began. They looked at the forms again to check their consistency. They compared the data with Social Security Administration (SSA) files. They looked at Medicare enrollment files and samples of death certificates. They concluded that the United States needed a complete registration system for 100 years before it was likely that the true number of centenarians could be known. Even then, we should not expect precision.

In kindergarten, we count with fingers or beads; in life we are faced with jam. And all this at the relatively simple end of counting, where definitions are clear, the subjects have a real presence, and everyone knows what they are looking for. Whether you are aged 100 or more ought to be a matter of simple fact. But if the simple, well-defined facts are not always easy to come by, what happens when the task becomes just a little more complicated?

"Yob Britain! 1 in 4 Teen Boys Is a Criminal!" said headlines in January 2005. "1 in 4 teen boys claims they have done a robbery, burglary, assault or sold drugs." As another newspaper put it: "Welcome to Yob UK!"

A few years ago, the National Institute on Alcohol Abuse and Alcoholism achieved similar excitement in the United States. The headline in that case was just as dramatic, describing what seemed a booze-fueled punch-up of epidemic proportions, even among those nice kids who go to college, blaming alcohol for "600,000 assaults on campus" each year.

The survey of teenage boys in Britain suggested the old country had nurtured a breed of thugs, thieves, and pushers. Newspapers mourned parenting or civilization, politicians held their heads in their hands in woe, and the survey itself really did say that 1 in 4 teenage boys was a "prolific or serious offender."

A competent survey asking boys what they'd been up to and then counting the answers ought, we're all tempted to think, to be straightforward (assuming the boys told the truth). Counting, after all, is simple enough for kindergarten, where one number follows another, each distinct and all units consistent: 1, 2, 3 . . .

The common, unconscious assumption is that this child's-play model of counting still applies, that it is a model with iron clarity in which numbers tick over like clockwork to reach their total. But when counting anything that matters in our social or political world, although we act as if the simple rules apply, they do not, they cannot, and to behave otherwise is to indulge a childish fantasy of orderliness in a world of windblown adult jumble.

It is, for a start, a fundamental of almost any statistic that, in order to produce it, something somewhere has been defined and identified. Never underestimate how much nuisance that small practical detail can cause.

First, it has to be agreed what to count. What is so hard about that? Take the laughably simple example of three sheep in a field:

What do we want to count?

Sheep.

How many are there?

Three.

But one is a lamb. Does that still count as a sheep? Or is it half a sheep? One is a pregnant ewe in advanced labor. Is that one sheep, or two, or one and a half? (We assume no multiple births.) So what is the total? Depending on how the units are defined, the total could be 2, 2.5, 3, 3.5, or 4. For a small number, that's quite a spread; one answer is twice the size of another, and counting to four just became ridiculous. In the real world, it often does whenever the lazy use of numbers belittles the everyday muddle.

That's counting sheep. Think what must go on, for example, in government statistics aiming to capture changes in the lives of millions of people, prices, and decisions. Take the number out of work. When do we call someone unemployed? Must they be entirely unemployed or can they do some work and still qualify? How much? An hour a week, or two, or ten? Do they have to look actively for work or just be without it? If they do have to look, how hard? What if they work a bit, but they are not paid, as volunteers? Mrs. Thatcher's Conservative governments notoriously changed the definition of "unemployed" twenty-three times (or was it twenty-seven—there is some disagreement!).

Numbers, pure and precise in abstract, lose precision in the real world. It is as if they are two different substances. In math they seem hard, pristine, and bright, neatly defined around the edges. In life, we do better to think of something murkier and softer. It is the difference, in one sense, between diamonds and a thick strawberry jam; hard to believe it is a difference we forget or neglect. Too often counting becomes an exercise in suppressing life's imprecision.

In that real world of soft edges, what was the definition used to identify those drunken assaults on campus, or the teenage boys as serious or prolific offenders? The largest category of offense in the UK example, by far, was assault, and judged serious if it caused injury. Here is how it was defined by a survey question:

Have you ever used force or violence on someone on purpose, for example by scratching, hitting, kicking, or throwing things, which you think injured them in some way?

And then, deliciously:

Please include your family and people you know as well as strangers.

Fifty-eight percent of assaults turn out to have been "pushing" or "grabbing." Thirty-six percent were against siblings. So anyone who pushes a brother or sister six times leaving them no worse for wear is counted a "prolific offender." Push little brother or sister even once and bruise an arm and you are a "serious offender," since the offense led to injury. Or, as the press put it, you are a yob, a thug, bracketed with drug dealers, burglars, murderers, and every other extremity of juvenile psychopath.

The characteristics of these surveys mirror the alarmed reporting of findings of bullying in U.S. schools, discussed by Joel Best in his book *More Damned Lies and Statistics*. Here the headline claim was that 30 percent of children in grades 6 through 10 had moderate or frequent involvement in bullying. A big number, reached by defining "involvement" as including both bullies and those they bullied, and by defining "moderate bullying" as happening "sometimes." And the definition of bullying? Joel Best looked one up from the Justice Department:

"Bullying can take three forms: physical (hitting, kicking, spitting, pushing, taking personal belongings); verbal (taunting, malicious teasing, name calling, making threats); and psychological (spreading rumors, manipulating social relationships, or engaging in social exclusion, extortion, or intimidation)."

As Best has remarked: "We're talking about people in grades 6 to 10. Does anyone remember high school?"

In the U.S. survey of drinking and assault, the numbers "on campus" were similarly stretched by also including anyone who had enrolled in a college course after work for one night a week. According to the U.S. Census Bureau, there are about 18 million college students enrolled in college in the United States, about 5 million of them part-time.

The survey question asked students whether they had been "assaulted, pushed, or hit" "because of other students' drinking" at some time during the school year. Does that include pushing in the line at the bar? Alcohol is certainly a factor in some ugly crime, including among students, but whether this survey conveyed the real extent, we wouldn't like to say.

Each time we count something, we define it; we say the things we count are sufficiently the same to be lumped together. But most of the important things we want to count are messy, like people, behaving in odd ways, subtly and not so subtly different. They do not stand still, they change, their circumstances vary widely. So how do we nail them down in one category under one set of numbers? Using what definitions? Until this context is clear, how can it even be said what has actually been counted?

Behaviors that seem clear-cut and therefore easily countable at the extremes often blur into one another somewhere along a line of severity. Sister-shovers and knife-wielding maniacs differ, so why are they lumped together in the reporting of the British yob survey? In determining a definition of what to count, the researchers looked to the law, whose long arm can indeed be brought to bear on pushing, shoving, etc.; can be, but normally isn't. The wise cop on the beat who decides that the brothers with grazed elbows from No. 42 are squabbling kids, not criminals fit for prosecution, is making a human judgment with which formal definition struggles. Definition does not like discretion and so tempts us to count rigidly, as teacher taught, in this case by saying that behavior is either unlawful or not. But this fancifully optimistic classification, insisting on false clarity, serves only to produce a deceptive number. It is astonishing how often the "1 in 4" headline comes along for one social phenomenon or another. It is invariably far tidier than it ought to be, often to the point of absurdity. The paradox of this kind of neatness is that it obscures; it is a false clarity leading to a warped perspective, and an occupational hazard of paying insufficiently inquiring attention to the news, in numbers above all.

Some of the crime identified by the survey certainly was brutal but, given the definitions, what were the headline numbers from the survey really counting? Was there clear statistical evidence of Yob

UK? Or did the survey rather reveal a remarkably bovine placidity in Britain's sons and brothers? After all, 75 percent of all teenage boys claimed not to have pushed, grabbed, scratched, or kicked as many as six times in the past year (what, not even in the school dinner line?), nor to have done any of these things once in a way that caused even the mildest scrape. According to these definitions, the authors of this book must make their confession: both were serious or prolific offenders in their youth. Reader, your writers were yobs. But perhaps you, too, were one of us?

We make no argument about whether the behavior of American or British youth is better or worse than it used to be or about trends in American student drinking. We note examples of appalling behavior like everyone else. We do argue that the headline statistics—reported in a way that implied things were getting worse—were scant evidence for it, being scant evidence of anything very much. Numbers, claiming with confidence to have counted something breathtaking, mug our attention on every corner. These surveys sometimes produce interesting details, if we can be bothered to read them properly, but the summary statistics often count for nothing.

It wouldn't sell, but a more accurate tabloid headline might have read: "One in four boys, depending how they interpret our question, admits getting up to something or other that isn't nice, a bit thoughtless maybe, and is sometimes truly vicious and nasty; more than that, we can't really say on the evidence available to us."

Let us assume the details are usually worked out properly, that we agree whether we mean sheep only or sheep *and* lambs, *and* we say so in advance, *and* we've decided on a clear way of telling the difference, and that what reports of the survey meant to say was that if you look at enough boys you will find a range from brutal criminal thug, through bully, all the way to occasionally unthinking or slightly boisterous lad, now and then, in the dinner line, maybe. Less dramatic, less novel, and oh, all right, you knew it already; but at least that would be a little more accurate. Sticking a number onto trends in juvenile behavior gives an air of false precision, an image of dia-mond fact, and in this case forces life into categories too narrow to contain it. Picture that offending more properly as a vast vat of straw-berry jam, think again about what single number can represent it, and see this massively difficult measurement for what it is. The fault here is not with numbers and the inevitable way that they must bully reality into some semblance of orderliness. It is with people, and our tendency to ignore that this compromise took place, while leaping to big conclusions.

Is this a mere tabloid extravagance? Not at all: it is common-place, in policymaking circles as in the media.

When, amid fears of a pension crisis, the British government–appointed Turner Commission published a preliminary report in 2005 on the dry business of pension reform, it said 40 percent of the population was heading for "inadequate" provision in retirement. With luck, your definitional muscles will now be flexing: what do they mean by "inadequate"?

In reaching that 40 percent figure—a shocking one—the Com-mission had said each to their own resources: either you have a pen-sion or you don't, a hard-and-fast definition like the interpretation in the yob survey of the law on assault, in or out but nothing in be-tween, covered or not, according to your own and only your own finances.

But perhaps you married young, perhaps you never had a paid job, perhaps your partner earned or still earns a sum equal to the GDP of a small state, perhaps you had been together for forty years, living in a château. All things considered, you might expect a comfortable sunset to life. But if you were fool enough in the Commission's eyes to have no savings or pension in your own name—shame on you—you were counted as lacking adequate provision for retirement.

This was a figure that mattered greatly to policymakers, since it seemed to suggest on an epic scale an irrational population in denial about the possibility of old age. Marriage—and the share of pensions that might be expected to go with it (hardly an unreasonable detail)—was one of many definitional question marks over that assessment. Soon afterward another report, an independent one, recalculated the number it believed to be without adequate pensions, this time taking marriage into account. It came up with 11 percent with inadequate savings instead of 40 percent. In later reports by the Pensions Commission itself, the 40 percent claim disappeared.

For the time being, we can establish a simple rule: if it has been counted, it has been defined, and that will often have meant using force to squeeze reality into boxes that don't fit. This is ancient knowledge, eternally neglected. In Book I of Aristotle's *Nicomachean Ethics* he writes, "It is the mark of an educated man to look for precision in each class of things just so far as the nature of the subject admits." But not more—and that is the critical point. Always think of the limitations. Always ask: "Are the definitions diamond-hard or strawberry jam?" "Either way, am I content with the units they define?" In short: 1, 2, 3 of what? And are there really 4, 5, and 6 of the same?

SIZE 2

IT'S PERSONAL

Simplify numbers and they become clear; clarify numbers and you stand apart with rare authority. So begin in the simplest way possible, with a question whose wide-eyed innocence defies belief:

"Is that a *big* number?"

Do not be put off by its naïveté. The question is no joke. It may sound trivial, but it captures the most obstinate problem with the way numbers are produced and consumed. Zeros on the end of a number are often flaunted with bravado, to impress or alarm, but on their own mean nothing. Political animals especially fear the size question, since their dirty secret is that they seldom know the answer. A keen sense of proportion is the first skill—and the one most bizarrely neglected—for seeing through numbers. Fortunately, everyone can do it. Often all they have to do is think for themselves.

The Danish economy has been in serious trouble. According to news broadcast by National Public Radio in 2005, a decision by a Danish newspaper to publish satirical cartoons of the Prophet Muhammed led to catastrophic losses by one Danish company of $200 million a day on its Middle East operations.

$200 million—is that a *big* number?

Daniel Adderley is aged nearly sixty-seven. He is lucky to be alive, according to the British newspaper the *Daily Telegraph*; the past two years living near the small town of Dorking in Southern England have been more risky for him than for soldiers on the front line in Afghanistan or Iraq. Not that Dorking is unusually perilous, in case you are wondering, simply that in November 2005 a front-page story in the *Daily Telegraph* reported government plans to raise the age of retirement for men from sixty-five to sixty-seven. If enacted, the paper said, one in five who formerly would have survived long enough to collect a pension would now die before receiving a penny. Hundreds of thousands would be denied by two cruel years.

1 in 5—is that a *big* number?

In 1997 the British Labour government said it would spend an extra £300 million (about $600 million) over five years to create a million new child care places.

£300 million. Is that a *big* number?

In 2006 the British National Health Service was found to be heading for a budget deficit of nearly £1 billion (about $2 billion).

£1 billion. Is that a *big* number?

The answer to the first question is that yes, $200 million a day is an enormous number for a small country like Denmark. It would hurt badly enough in the United States. The total output of the whole Danish economy (Danish GDP) works out at approximately $600 million a day. So losses by one company of $200 million a day are equal to about one third of the entire output of the entire economy. Is it at all likely that one company, in its operations in one part of the world, could account for so much? Possibly, if it was betting the national reserves at the roulette wheel.

The true figure was said to be about $2 million a day, still a substantial loss for the firm concerned, no doubt, but not quite surpassing the scale of the Great Depression.

Next, Daniel Adderley's chances of dying in Dorking and the effect of a change in pensions. The answer to whether 1 in 5 is a big number is yes, 1 in 5 men aged sixty-five would be a catastrophic number to die in two years, a number to strike terror into every sixty-five-year-old, a number so grotesquely enormous that some people at the *Telegraph* surely should have asked themselves: Could it be true? Perhaps, if the plague returned, otherwise it doesn't take much thinking to see that the report was ridiculous. But it is a simple sort of thinking even smart journalists often do not do.

According to Britain's National Statistics (www.statistics.gov.uk), about 4 percent of sixty-five-year-old men die in the following two years, not 20 percent. About 20 percent of those born do indeed die before they reach the age of sixty-seven, but not *between* the ages of sixty-five and sixty-seven. Misreading a number in a table, as the journalists seem to have done, is forgivable; failure to ask whether the figure makes the sort of sense they could see every day with their own eyes is less so. For the report to be correct, more than 100,000 more men aged sixty-five and sixty-six would have to die every two years. The equivalent figure for the United States would be about 500,000. They would be turning up their toes all over the golf course. All things considered, we think Daniel will make his sixty-eighth birthday.

Next, Labour's £300 million for child care. Here, no one involved in the public argument, neither media nor politicians, seemed to doubt its vastness. The only terms in which the opposition challenged the policy were over the wisdom of blowing so much public money on a meddlesome idea.

So is £300 million to provide a million places a big number? Share it out and it equals £300 per place. Divide it by five to find its worth in any one year (remember, it was spread over five years), and you are left with £60 per year. Spread that across fifty-two weeks of the year and it leaves £1.15 per week. Could you find child care for £1.15 a week? In parts of rural China, maybe. Britain's entire political and media classes discussed the policy as if you could.

But by "create" the government must have meant something other than "pay for" (though we wonder if it minded being misunderstood), something perhaps along the lines of "throw in a small sum in order to persuade other people to pay for . . ." And yet the coverage oozed a sense of bonanza.

Does the public debate really not know what "big" is? Apparently not, nor does it seem to care that it doesn't know. When we asked the head of one of Britain's largest news organizations why journalists had not spotted the absurdity, he acknowledged there was an absurdity to spot, but said he wasn't sure that was their job. For the rest of us, to outshine all this is preposterously easy. For a start, next time someone uses a number, do not assume they have asked themselves even the simplest question. Can such an absurdly simple question be the key to numbers and the policies reliant on them? Often, it can.

The fourth example, the £1 billion National Health Service (NHS) deficit, was roundly condemned as a mark of crisis and mismanagement, perhaps the beginning of the end for the last great throw of taxpayers' money to prove that a state system worked. But was it a *big* number?

The projected deficit had fallen to about £800 million at the time of writing, or about 1 percent of the health service budget. But if that makes the NHS bad, what of the rest of the government? The average Treasury error when forecasting the government budget deficit one year ahead is 2 percent of total government spending. In other words, the NHS, in one of its darkest hours, performed about twice as well against its budgetary target as the government as a whole. There are few large businesses that would think hitting a financial target to within 1 percent anything other than management of magical precision. They would laugh at the thought of drastic remedial action over so small a sum, knowing that to achieve better would require luck or a clairvoyant. NHS spending is equivalent to about £1,600 per head, per year (in 2007), of which 1 percent is £16 (about $32), or less than the cost of one visit to a GP (about £18). No doubt there was mismanagement in the NHS in 2006 and, within the total deficit, there were big variations between the individual NHS trusts, some of which faced a genuine problem, but given the immense size of that organization, was it of crisis proportions, dooming the entirety? It is often true that the most important question to ask about a number—and you would be amazed how infrequently people do—is the simplest.

Each time a reporter or politician turns emphatic and earnest, propped up on the rock of hard numbers like the bar of the local, telling of thousands, or millions, or billions of this or that, spent, done, cut, lost, up, down, affected, improved, added, saved . . . it is worth asking, in all innocence: "Is that a *big* number?"

In this chapter, six will be big and a trillion won't, without resorting to astronomy for our examples, but taking them from ordinary human experience. The human scale is often forgotten when size gets out of hand, and yet this is the surest tool for making numbers meaningful. Nor is there anything the least bit difficult about it. For it is also a scale with which we all come personally equipped.

For example, we asked people to tell us from a list of multiple-choice answers roughly how much the government spent on the health service in the UK that year (2005). The options ranged from £7 million (about $14 million) to £7 trillion (about $14 trillion); so did people's answers. Since being wrong meant under- or overestimating the true figure by a factor of at least 10 and up to 10,000, it is depressing how wrong many were. The correct figure at that time was about £70 billion (70,000,000,000).

Some people find anything with an "-illion" on the end incomprehensible; you sense the mental fuses blowing at anything above the size of a typical annual salary or the more everyday mortgage. A total NHS budget of £7 million is about the price of a large house in certain exclusive parts of London (and so somewhat unlikely to stretch to a new hospital), or equivalent to health spending per head of about 12 pence a year. How much health care could you buy for 12 pence a year? A total NHS budget of £7 trillion would have been more than six times as big as the whole economy, or about £120,000 of NHS spending a year for every single member of the population. When we told people that their guess of £7 million implied annual health spending per head of about 12 pence, some opted instead for a radical increase to . . . £70 million (or £1.20 per head). How many trips to the GP would that pay for? How many heart transplants? These people have forgotten how big they are. Put aside economics, all that is needed here is a sense each of us already has, if only we would use it: a sense of our own proportions.

Millions, billions . . . if they all sound like the same thing, a big blurred thing on the evening news, it is perhaps because they lack a sense of relative size that works on a human scale. So one useful trick is to imagine those numbers instead as seconds. A million seconds is about 11.5 days. A billion seconds is nearly thirty-two years.

If £300 million can be pitifully small, and 1 in 5 outrageously big, how do we know what's big and what's small? (One of these is a number and the other a ratio, but both measure quantity.)

The first point to get out of the way is that the amount of zeros on the end of a number gets us nowhere; that much will be painfully obvious to many, even if some choose for their own reasons to ignore it. Immunity to being impressed by the words billion or million is a precondition for making sense of numbers in public argument. In politics and economics, almost all numbers come trailing zeros for the simple reason that there is a teeming mass of us and oodles of money in the economy of an entire country that produces more than $14,000,000,000,000 annually, as the United States does, with a population of 300,000,000 people. Plentiful zeros come with the territory. Our default position should be that no number, not a single one of them, regardless of its bluster, is either big or small until we know more about it.

All this suggests an easy solution, already hinted at; that the best way to see through a number is to share it out and make it properly our own. Throwing away the mental shortcut—"lots of zeros = big"—forces us to do a small calculation: divide the big number by all the people it is supposed to affect. Often that makes it strangely humble and manageable, cutting it to a size, after all, that would mean something to any individual, so that anyone can ask: "Is it still big if I look at my share of it?"

Well, could you buy child care for £1.15 a week? That is a judgment you can easily make. Will £300 million pay for a million extra child care places for five years? That sounds harder, though it is much the same question. Making the hard question easy is not remotely technically difficult; it is mainly a matter of having the confidence or imagination to do it.

Maybe on hearing numbers relating to the whole nation, some fail to make the personal adjustment properly: Here's little me on average earnings and there's £300 million. But it is not all yours. To make a number personal, it has to be converted to a personal scale, not simply compared with the typical bank balance. The mistake is a little like watching the teacher arrive with a big bag of sweets for the class and failing to realize that it means only one each. Yet being taken in by a number with a bit of swagger is a mistake that is often made. The largest pie can still be too small if each person's share is a crumb.

A convenient number to help in this sort of calculation is 15.6 billion (15,600,000,000), which is the U.S. population (300,000,000), multiplied by fifty-two, the number of weeks in a year. This is about how much the government needs to spend in a year to equal $1 per person per week in the United States. Divide any public spending announcement by 15.6 billion to see its weekly worth if shared out among us.

Some readers will be aghast that any of this needs saying, but the need is transparent and those in authority most often to blame. Part of our contention is that salvation from much of the nonsense out there is often straightforward, sometimes laughably so, and the simpler the remedy the more scandalous the need. Of course, not all cases are so routine, not all numbers need sharing equally, for example, often being aimed more at some than others. That's another simple but important principle to which we will return. For now it serves to emphasize the point that we can make more sense of a number if we first share it out where it is due.

None of this is to encourage cynicism; it is about how to know better. And it is worth saying that innumeracy is not the same as dishonesty. Numbers do grow because people try to pull the wool over our eyes, sure enough, but also because they are themselves muddled, or so eager for evidence for their cause that they forgo plausibility. Maybe the *Daily Telegraph* journalists, writing about dying pensioners, were so taken with the notion that Prime Minister Blair and company were beastly to old folk who had slaved for an honest crust, that they allowed this apparently satisfying notion to make mush of their numeracy.

This tendency of big brains to become addled over size is why the peddlers of numbers often know less about them than their audience, namely, not much; and why questions are always legitimate, however simple. Size matters. It is odd trying to persuade people to give it more attention, but this is attention of a neglected kind, where, instead of simply bowing to a number with apparent size and attitude, we insist on a human-scale tape measure.

Size is especially neglected in the special case when mixed with fear. Here, there is often no need even to claim any great magnitude; the simple existence of danger is enough, in any quantity.

To see this, and the kind of brainstorm that talks of toxicity or poison as if it means invariably fatal by the micron, think of an impact to the head. If delivered by a kiss to your child's brow, it is seldom harmful; if by a falling girder from the rooftop, it is probably as serious as harm can be. It is clear that impacts to the head have variable consequences depending, well, on the force of the impact, and that below a certain threshold, they do no harm whatsoever. Readers of this book asked to endure an impact to the head would, we hope, not be alone in asking, "How hard?"

This humdrum principle—that harm depends on how much of a given danger people are exposed to, and that at low levels it often implies no harm at all—is one that everyone applies unthinkingly umpteen times a day. Except to toxicity. Food and environmental health scares are a paranoia-laced anomaly, often reported and, presumably, often read with no sense of proportion whatsoever, headlined with statements the equivalent of "Impact to Head Fatal, Says Study."

Let's labor the metaphor to bring out its relevance. An impact to the head has variable consequences, depending on the dose. So what happens to this elementary principle when, as happened in 2005, someone tells us that cooked potatoes contain a toxic substance called acrylamide?

Acrylamide is used industrially—and also produced by a combination of sugars and amino acids during cooking—and known at some doses to cause cancer in animals and nerve damage in humans. The scare originated in Sweden in 1997 after cows drinking water heavily polluted with industrial acrylamide were found staggering around as if with mad cow disease. Further Swedish research, the results of which were announced in 2002, found acrylamide in a wide variety of cooked foods.

On average, people are reckoned to consume less than 1 microgram of acrylamide per kilogram of body weight per day (though studies vary). That is one one thousandth of the dose associated with a small increase in cancer in rats. Some will consume more, though in almost all epidemiological studies have seemed to get no more cancer than anyone else.

Salt is often added to potatoes, particularly fries and chips, liberally. It is both essential for survival and also so toxic it can kill a baby with smaller quantities than may be contained in a salt shaker. About 3,750 milligrams per kilo of body weight is the accepted lethal dose (LD) for salt (the quantity that has been found to kill 50 percent of rats in a test group, known as the LD50). For a 3 kilogram baby, that equals about 11 grams, or a little more than 2 teaspoons. Does anyone report that 2 teaspoons of salt sprinkled on your fries flirts with death? What is said, for a variety of health reasons, is that we should keep a casual eye on the quantity we consume. Be sensible is the unexcitable advice. Even a bit of arsenic in the diet is apparently essential for good health.

So why such absolutism about the quantities of acrylamide? Maybe the dual advantages of health paranoia and never having heard of the stuff before made panic more inviting. Had journalists thought about the numbers involved, it might have kept fear in proportion: The quantity of acrylamide equivalent to that associated with a small increased risk of cancer in rats turned out, according to one estimate, to require consumption of about 65 pounds of cooked potatoes (about a third to half of the typical human body weight) every day for years.

When the headline reveals toxicity, the wise reader, aware that most things are toxic at some dose, asks, "In what proportions?" Water keeps you alive, unless, as thirsty consumers of the drug ecstasy are often told, you drink too much, when it can cause hyponaetraemia (water poisoning). Leah Betts, a schoolgirl, died after taking ecstasy then drinking fourteen pints of water in ninety minutes.

That is why all toxicologists learn the same mantra from their first textbooks, that toxicity is in the dose. This does not mean all claims of a risk from toxicity can be ridiculed, simply that we should encourage the same test we apply elsewhere, the test of relevant human proportion. Keep asking the wonderful question, with your-self in mind: "How big is it?"

There are exceptions: peanut allergies, for example, can be fatal at such small doses that it would be unhelpful to say the sufferer must quibble over the quantity. Otherwise, though treated like one, "toxic" is not necessarily a synonym for "panic!"

Another case where size receives less than its due is the reporting of genetics. Here, headlines typically tell us that researchers have dis-covered "the gene for" or "genetic basis of" or "gene linked with." The tendency is to talk of these genes as if they revealed the secret one-way sign of life. At the very least, we assume that genes must give us the heaviest of hints.

With some conditions—cystic fibrosis is an example—this is an apt characterization of the truth. If you have the genetic mutation, you have cystic fibrosis; if you don't, you don't. It is a real example of 100 percent genetic determinism.

But it is not typical. More commonly, all that the discovery of "the gene for" really tells us is that a gene is present in more of the people who have a condition than in those who don't. That is, it is not a one-way sign but no more than encouragement to take this turn rather than that. This is what makes the subject, often as not, a matter of size. For the critical question, almost never answered clearly in news coverage, is "how much encouragement"?

For example, if you had to guess how many people with multiple sclerosis have "the gene for" MS, you might be tempted to answer "all of them" or else how could they have the condition? In fact, a study identifying the "genes for" MS (two culprits were found) showed that one of these genes is present in 87 percent of people who have MS. But 13 percent of those who have MS *do not* have the "MS gene."

Asked how many have "the gene for MS" but do not have the condition, you might reason that the gene is a predisposition, not a sure thing, and so a lucky few will avoid MS despite having the gene for it. In fact, that same study found the gene in 85 percent of people who never develop MS.

The other "gene for" MS is present in about 78 percent of those who have MS and 75 percent of those who don't. These differences are real, as far as researchers can tell, but tiny. For any individual, the chance of developing multiple sclerosis is almost unaltered whether they have the genes or not, namely, extremely small in both cases. MS is thought to have a prevalence of between about 100 and 200 per 100,000, depending where you live. It is a rare illness and its prevalence is only minutely increased among those who have the genes "for" it.

Restless leg syndrome is another recently identified "gene for." Three studies in three countries found that a gene seemed to be in part responsible. How big a part? One study found that it was present in 83 percent of sufferers and 76 percent of non-sufferers. The other study said 76 percent and 68 percent respectively. The third 77 percent and 65 percent. Again, the differences are smaller than we might expect. That is, the gene plays a part, though probably only a small one. As one critic said, it might be more true to say that the gene has something to do with something else to do with something related to restless leg syndrome.

It is a similar story with asthma. A recently discovered "gene for" asthma is present in about 62 percent of sufferers and about 52 percent of the asymptomatic population. For any individual, having the gene makes little difference to the chance of having asthma.

Some genes are more influential. With some conditions, a group of genes is implicated, each one having a tiny effect on the probabilities, but cumulatively making a more appreciable difference. But reading breathless reports of new genetic discoveries, how many have a sense of the proportions involved? How many, we wonder, have a sense that proportion is a relevant consideration? Playing nature against nurture is a great game. Biology, however, is not often black or white. But nor would it be hard to convey a sense of the shades of gray.

By now you might be all but convinced of the merit of our coy little question, convinced that size matters but is often untested, and that long lines of zeros tell us nothing. So let's measure that assurance with a real monster of a number: £1,000,000,000,000—or 1 trillion pounds (about $2,000,000,000,000), as we now call such a quantity of zeros, in accordance with international economic convention. Imagine this was what British people owed, the total of their debts. Do not be intimidated by the apparent size—go ahead and ask: "Is that a *big* number?"

Most newspapers thought so, when we reached that figure in 2006, and some splashed it across the front page. The answer is that it is highly debatable whether debts of £1 trillion today are large or not. It was the highest figure ever, true, since surpassed, but we can also say with supreme confidence that it will be surpassed again, and again, since records are all but inevitable in a growing economy. The reporting implied surprise; but the combination of inflation and a growing economy means that the total number of pounds in Britain rises by about 5 percent a year. This doubles the number of pounds in the economy every fifteen years. When there are so many more pounds of everything, is it any surprise there are also more pounds of debt?

So put aside the manufactured sense of shock at this number and try pitching at it one of our innocent little questions: How is it shared out? Not evenly, is the unsurprising answer. Nor is it the prototypical shopaholic with a dozen maxed-out credit cards who accounts for any but a tiny part of the trillion. It is, in fact, the rich who owe overwhelmingly the biggest proportion of the debt, and always have, often in mortgages taken out to pay for houses that are growing in value, which means they also have increasing assets to repay their debts if they have to.

To see the oddity of talking about debt in the tone often employed, apply it to your own debts and see how you fare. First, how much did you owe at age fifteen? Four dollars and twenty cents—to your brother—how prudent. How much when you were of working age? What? It went up? Presumably this left you much worse off, at the beginning of a long and shocking slide into middle-age profligacy, perhaps. And if you subsequently took out a mortgage, why, your debts probably reached *a record*! This must have been, obviously, the most miserable, destitute moment of your life, as you pined for the financial circumstances of a fifteen-year-old once more.

This is—probably—not the best way to characterize debt, even if it is in a hallowed media tradition. Is larger debt mostly a reflection of larger borrowing power, linked to a rising ability to sustain higher repayments? Could it be that increased debt, far from being a sure measure of disaster, is, in many cases, a sign that life is getting better? This is the typical personal experience and it makes the national picture eminently predictable. Debt is a corrosive problem for those who cannot afford it, and there do genuinely seem to be more people in that position, but this has precious little to do with the trillion pounds, most of which is a sign not of penury, but wealth. While it is true that debts at the extreme can become a serious problem for governments, companies, or individuals, it is also true that they can be a sign of robust economic good health. Some might think it madness to make this argument after the "credit crunch" of 2007–2008. But that episode was a result of bad lending and borrowing decisions and, in the United States in particular, money lent to people who were unlikely to be able to repay. Such foolishness on the part of large financial institutions is remarkable, and damaging, but is not evidence that borrowing is inherently bad, rather that irresponsible lending is bad.

Another way of thinking about this is to say that your debts would be too big only if you could not repay them. But how many news reports of a nation's debts also report its wealth? Quite absurdly, it almost never gets a mention, even though it would be one of everyone's first considerations. We have taken to calling this the Enron School of Reporting. The energy firm Enron, readers may recall, trumpeted its assets, but conveniently neglected to talk about its debts, and went spectacularly bust. News reporting about the public's financial state is often the other way around, blazing with headlines about debt, but neglecting to mention wealth or income. So trust what is known to be true for the individual, take the calculation anyone would make personally, and apply it to the national picture. This shows (see chart) that personal wealth has been increasing over recent years steadily and substantially. In 1987 the total personal wealth of all the households in the UK added up to about four times as much as the annual income of the whole country. By 2005 wealth was six times as great as annual national income.

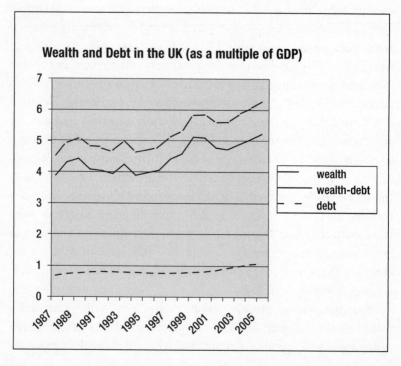

Wealth and Debt in the UK (as a multiple of GDP)

Not poorer, but prodigiously richer, is the story of the last twenty or so years. This wealth is held in houses, pensions, shares, bank and building society accounts, and it is held very unequally, with the rich holding most of the wealth, as well as most of the debt. Some of the increase reflects house-price growth, some increases in share prices, but there is no denying that as the economy has grown, and incomes with it, so have savings and wealth. Debt has been rising, but wealth has been rising much more quickly.

Even if we look at the category of debt said to be running most seriously out of control—unsecured personal debt (which includes credit cards)—we find that as a proportion of what we have available to spend—household disposable income—this debt has remained steady for the last five years (before which it did go up—a bit). Once more, it is unequally distributed and more of a problem for some than others, but for precisely those reasons it makes no sense to use the total number as if it says anything whatsoever about those in the most dire financial straits.

How can we know from personal experience that the debt number may not be as monstrous as it looks? By thinking about our own life cycle and our own borrowing. We are happier to borrow more as we grow richer, and often do, but it tends to become less rather than more of a problem. With that much clear, it is easy to see that rising debt might well indicate rising wealth in the population as a whole. Essentially, we use the same method as before, we share out the number where it belongs, rather than according to what makes a good story. So we should share it mostly among the rich, far less among the poor. Just for good measure, £1 trillion of debt shared out across the population is a little less than £17,000 per person, compared to wealth, on average, of over £100,000.

It is not only debt in the UK that can more usefully be put into human proportions. In the summer of 2005 at the Gleneagles summit—held during the UK's presidency of the G8 group of countries—the Chancellor of the Exchequer Gordon Brown announced that $50 billion of developing country debt would be written off. It sounds generous, and is important for developing countries, but how generous does it look from inside the G8? Is it a *big* number for us?

Fifty billion dollars was the stock of outstanding debt. By writing if off, what the G8 countries actually gave up were the annual repayments they would have received, often at preferential interest rates, equal to about $1.5 billion a year. This was the real cost, substantially lower than the headline figure. Convert $1.5 billion into pounds, and it seems smaller still, at around £800 million at the then prevailing exchange rates. £800 million a year still sounds like a lot, but remember the exhortation to make numbers personal. The population of the G8, conveniently, is around 800 million. So the Gleneagles deal will cost each person living in the G8 about £1 a year. And even that is an exaggeration, since much of the money will come out of government aid budgets that would have gone up anyway, but will now be diverted to pay for the Gleneagles deal, leaving the best estimate of how much more Gleneagles will cost each of us than had it never happened, as close to zero.

Let's look, finally, at the other end of the scale, to a number in single figures, a tiny number: six. You will be honing your skepticism by now, reluctant to take anyone's word or commit to an opinion, and wanting to know more of the context. That is as it should be. The six we have in mind is the celebrated six degrees of separation said to exist between any two people on the planet. So, is six a small number?

The idea was first proposed—according to Wikipedia—in 1922 in a short story titled "Chains" by a Hungarian writer, Frigyes Karinthy. But it was an American sociologist, Stanley Milgram, who became most closely associated with the suggestion when, in 1967, he claimed to have demonstrated empirically that six steps was generally more than enough to connect any two people in the United States. He called this "the small world phenomenon."

Milgram recruited nearly 300 volunteers whom he called "starters," asking each to send a gold-embossed passportlike package to a stranger, usually, but not always, in another city. They could use any intermediary they knew on first-name terms, who would then try to find another who could do the same, each time moving closer to the target. All they had to go on was the recipient's name, home town, occupation, and a few descriptive personal details, but no address. Milgram reported that 80 percent of packages reached their target in no more than four steps and almost all in fewer than six.

The story became legendary. Then, in 2001, Judith Kleinfeld, a psychologist at the University of Alaska, became interested in the phenomenon and went to study Milgram's research notes. What she found, she says, was disconcerting: first, that an initial unreported study had a success rate of less than 5 percent. Then, that in the main study more than 70 percent of packages had never reached their intended destination, a failure rate that raises doubts, she says, about the whole claim. "It *might* be true," she told us, "but would you say that the packages arrived in fewer than six steps when 70 percent never arrived at all?" Furthermore, in criticism that bears more closely on our question, she notes that senders and recipients were of the same social class and above-average incomes, and all likely to be well connected.

So six may or may not be the right number, and the central claim in the experiment has never been satisfactorily replicated, but would it, if true, nevertheless be a small number? The point that connections are easier among similar people is a clue, and a prompt. This encourages us to think that not only the number of steps, but also the size of the steps matters. And if each step is giant, six will amount to an impossible magnitude.

Other studies, including more by Milgram himself, according to Judith Kleinfield's summary of his personal archives, found that where the packages crossed racial differences, the completion rate was 13 percent, rising to 30 percent in a study of starters and targets living in the same urban area. When, in another study, the step was from a low-income starter to a high-income target, the completion rate appears to have been zero. Connections were not nearly so easy across unexceptional social strata.

And as Judith Kleinfeld herself points out: "A mother on welfare might be connected to the president of the United States by a chain of fewer than six degrees: Her caseworker might be on first name terms with her department head, who may know the mayor of Chicago, who may know the president of the United States. But does this mean anything from the perspective of the welfare mother? . . . We are used to thinking of 'six' as a small number, but in terms of spinning social worlds, in a practical sense, 'six' may be a large number indeed."

"Six" is usually small; a billion is usually large. But easy assumptions will not do when it comes to assessing size. We need to check them with relevant human proportion. Six steps to the president sounds quick, but let people try to take them. They can be an ocean apart and represent a world of difference. A billion dollars across the United States can be loose change, 6 cents each per week. We need to think, just a little, and to make sure the number has been properly converted to a human scale that recognizes human experience. Only then, but to powerful effect, can we use that personal benchmark. The best prompt to thinking is to ask the question that at least checks our presumptions, simple-headed though it may sound: "Is that a *big* number?"

CHANCE 3

THE TIGER THAT ISN'T

We think we know what chance looks like, expecting the numbers she wears to be a mess, haphazard, jumbled. Not so. Chance has a genius for disguise and fools us time and again. Frequently and entirely by accident, she appears in numbers that seem significant, orderly, coordinated, or slip into a pattern. People feel an overwhelming temptation to find meaning in these spectral hints that there is more to what they see than chance alone, like zealous detectives over-alert to explanation, and to dismiss with scorn the real probability: "It couldn't happen by chance!"

Sometimes, though more seldom than we think, we are right. Often we are suckered, and the apparent order is no order, the meaning no meaning, it merely resembles one. The upshot is that discovery after insight after revelation, all claiming to be dressed in the compelling evidence of numbers, will mean nothing. It was chance that draped them with illusion. The best way to avoid being taken in is to remind ourselves what chance is capable of. The short answer for many of us is: a lot more than we think.

Who was it, furtive, destructive, vengeful in the darkness? At around midnight on bonfire night—November 5, 2003—on the outskirts of the village of Wishaw in England, someone had crime in mind, convinced the cause was just.

Whoever it was—and no one has ever been charged—he, she, or they came with rope and haulage equipment. A few minutes later, on the narrow stretch between a livery and a field, the ten-year-old, seventy-five feet tall cell-phone mast on the outskirts of the village was first quietly unbolted, then brought crashing down. The signal from the mast ceased at 12:30 AM precisely. Police found no witnesses.

By morning, protestors surrounded the upended mast and re-fused to allow T-Mobile, its owners, to take it away or replace it. An attorney representing the protesters told the landowners they would not be permitted access because that meant crossing someone else's property. The protest quickly became a round-the-clock vigil with both sides paying private security companies to patrol the boundary.

The villagers' earnest objection had a despairing motivation. Since the mast had gone up, among the twenty households within about 500 yards, there had been nine cases of cancer. In their hearts, the reason seemed obvious. They were, they believed then and still believe now, a cancer cluster. How could such a thing happen by chance? How could so many cases in one place be explained except through the malign effect of powerful signals from that mast?

The villagers of Wishaw might be right. The mast has not been replaced and the strength of local feeling makes that unlikely, not now, perhaps not ever. And if there were a sudden increase in crime in Wishaw such that nine out of twenty households were burgled, they probably would be right to suspect a single cause. When two things happen at the same time, they are often related.

But not always, and if the villagers are wrong, the reason has to do with the strange ways of chance in large and complex systems. *If* they are wrong, it is most likely explained as a result of their inability—an inability most of us share—to accept that apparently unusual events happening simultaneously do not necessarily share the same cause, and that unusual patterns of numbers in life, including the incidence of illness, are not at all unusual, not necessarily due to some guiding force or single obvious culprit, but callously routine, normal, and sadly to be expected.

To see why, stand on the carpet—but choose one with a pile that is not too deep (you might in any case want a vacuum cleaner on hand)—take a bag of rice, pull the top of the packet wide open . . . and chuck the contents straight into the air. Your aim is to eject the whole lot skyward in one jolt. Let the rice rain down.

What you have done is create a chance distribution of rice grains over the carpet. Observe the way the rice is scattered. One thing the grains probably have not done is fall evenly. There are thin patches here, thicker ones there and, every so often, a much larger and distinct pile of rice: it has clustered.

Wherever cases of cancer bunch, people demand an explanation. With rice, they would see exactly the same sort of pattern, but does it need an explanation? Imagine each grain of rice as a cancer case falling across a map of the entire United States. Is it plausible that there will be an even distribution of cancer? The example shows that clustering, as the result of chance alone, is to be expected. The truly weird result would be if the rice had spread itself in a smooth, regular layer. Similarly, the genuinely odd pattern of illness would be an even spread of cases across the population.

We asked a computer to create a random pattern of dots for the next chart, similar to the rice effect. Imagine that it covers a partial

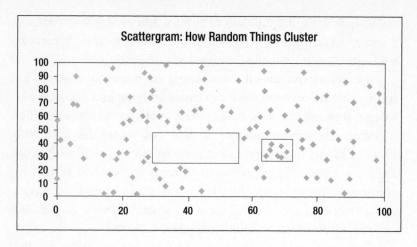

Scattergram: How Random Things Cluster

map of Michigan and shows cases of a rare cancer. One area, the square to the right (let's call it Ann Arbor), we might be tempted to describe as a cluster. The other, the rectangle, with no cases at all, might tempt us to speculate on some protective essence in the local water. The chart shows that such patterns can, and do, appear by chance.

It also illustrates a tendency known as the Texas Sharpshooter Fallacy. The alleged sharpshooter takes numerous shots at a barn (actually, he's a terrible shot—that's why it's a fallacy), then draws his bull's-eye afterward, around the holes that cluster.

This is sometimes what we do with cancer cases, saying "look how they all occurred here, in this small area!" As the Washington State Health department puts it: "In the same way, we might notice a number of cancer cases, then draw our population base around the smallest area possible, neglecting to remember that the cancer cases actually came from a much larger population."

The analogy draws no moral equivalence between cancer and rice patterns or pot shots, and people who have cancer have an entirely reasonable desire to know why. But the places in which cancer occurs sometimes will be clustered simply by obeying the same rules of chance as falling rice. They will not spread themselves uniformly across the country. They will also cluster, there will be some patches where cases are relatively few and far between and others with what seem worrying concentrations. Sometimes, though rarely, the worry will point to a shared local cause. More often the explanation lies in the often complicated and myriad causes of disease, mingled with the complicated and myriad influences on where we choose to live, combined with accidents of timing, all in a collision of endless possibilities which, somehow, just like the endless collisions of those rice grains in a maze of motion, come together in one place at one time to produce a cluster.

One confusion to overcome is the belief that the rice falls merely by chance, whereas illness always has a cause. This is a false distinction. "Chance" does not mean without cause—the position of the rice has cause all right; the cause of air currents, the force of your hand, the initial position of each grain in the packet, and it might be theoretically possible (though not in practice) to calculate the causes that lead some grains to cluster. Cancer in this respect—and only in this respect—is usually no different. For all the appearance of meaning, there is normally nothing but happenstance.

Think about the rice example for a moment beforehand and there is no problem predicting the outcome. Seeing the same effect on *people* is what makes it disconcerting. This is an odd double standard: Everyone knows things will, now and then, arrive in a bunch—it happens all the time—but in the event they feel put out; these happenings are inevitable, we know, yet such inevitabilities are labeled "mysterious," the normal is called "suspicious," and the predictable "perverse." Chance is an odd case where we must keep these instincts in check, and instead use past experience as a guide to what chance can do. We have seen surprising patterns before, often. We should believe our eyes and expect such patterns again, often.

People typically underestimate both the likely size of clusters and their frequency, as two quick experiments show. Shuffle a standard pack of fifty-two playing cards. Turn them from the top of the deck into two piles, red and black. How many would you expect in the longest run of one color or the other? The typical guess is about three. In fact, at least one run of five or six, of either red or black, is likely.

You can test expectations by asking a class of schoolchildren (say thirty people) to call in turn the result of an imaginary toss of a coin, repeated thirty times. What do they think will happen? Rob Eastaway, an author of popular math books, uses this trick in talks to schools, and says children typically feel a run of three of the same side will do before it's time to change. In fact, a run of five or more will be common. There will also be more runs of three or four than the children expect. Even though they know not to expect a uniform alternation between heads and tails on all thirty tosses, they still badly underestimate the power of chance to create something surprising.

But this is so surprising that we couldn't resist testing it: here are the real results of thirty tosses of a coin repeated three times (h = heads, t = tails). All clusters of four or more are in bold, and the longest are given in brackets.

1: *h t t h t t h h t h h* **t t t t t t** *h h* **h h** *t h h t t t h h*
    (*six tails in a row, four heads in a row*)
2: **t t t t** *h t t h h t h t* **h h h h** *t t t h h h t h h t h h t*
    (*four tails in a row, four heads in a row*)
3: *t h* **t t t t t** *h t t t h h* **t t t t t t** *h h h h t t h t h t h h t t*
    (*five tails in a row, six tails in a row*)

There was nothing obviously dodgy about the coin either: in the first test it fell fifteen heads and fifteen tails, in the second sixteen and fourteen, in the third ten and twenty, and the sequences are genuinely random.

So even in short runs there can be big clusters. The cruel conclusion applied to illness is that even if it were certain that phone masts had no effect whatsoever on health, even if they were all switched off permanently, we would still expect to find places like Wishaw with striking concentrations of cancers, purely by chance. This is hard to swallow: nine cases in twenty households—it must *mean* something. That thought encourages another: how can so much suffering be the result of chance alone? Who wouldn't bridle at an unfeeling universe acting at whim, where pure bad luck struck in such insane concentration? To submit to that fate would lay us open to intense vulnerability, and we can't have that.

But "chance" does not mean, in the ordinary meaning of these words, spread out, or shared, or messy. It does not mean what we would think of as disordered, or without the appearance of pattern. It does not, strictly speaking, even mean unpredictable, since we know such things will happen, we just don't know where or when; and chance certainly does not mean without cause. All those cancers had causes, but they are likely to have been many and various and their concentration no more than the chance result of many overlapping and unrelated causes. It is predictable that there will be cancer clusters in a country the size of the United States or UK, we just don't know where, and though it is a surprise when they appear, so is six consecutive tails in thirty tosses. Both ought to be expected.

Following a program in the *More or Less* series on BBC Radio 4 about randomness and cancer clusters, in which we nervously described to one of the Wishaw villagers active in a campaign against the phone mast how clusters can occur—a woman who herself had lived with cancer—we received an e-mail from a ferociously angry listener. How dare we, he said, take away their hope?

Chance is heartless. "As flies to wanton boys are we to the Gods," said Shakespeare's King Lear, "they kill us for their sport." More comforting amid hurt and distress, maybe, to find a culprit you can hope to hold accountable, or even destroy.

When the surgeon and academic Atul Gawande wrote in the late 1990s about why cancer clusters in the United States were seldom the real thing, he quoted the opinion of the chief of California's Division of Environmental and Occupational Disease Control that more than half the state's 5,000 districts (2,750 in fact) had a cancer rate above average. A moment's reflection tells us that this is more or less the result we should expect: simply put, if some are below average, others *must* be above, unless all are identical. With some grains spread out, others must be squeezed closer together. The health department in another state—Massachusetts—responded to between 3,000 and 4,000 cancer cluster alarms in one year alone, he said. Almost never substantiated, they nevertheless have to be investigated: the public anxiety is genuine and cannot be shrugged off, even if with the weary expectation of finding nothing. And sometimes, but rarely, they do find something. So these findings are not necessarily from reluctance by the medical or public health authorities to admit a problem. It is—usually—no conspiracy to suppress inconvenient facts, as the willingness to detect cancers in other ways shows. Unlike geographical clusters, discoveries of real occupational cancer clusters, or clusters of illness around a single life event— exposure to a drug or chemical for example—are not so unusual and have been readily divulged. The inconvenience of the malign effects of asbestos and tobacco has not prevented the medical authorities from exposing them, despite the resistance of those industries.

Even apart from those hooked by conspiracy theories, we are all gifted recognizers of patterns and the moments when a pattern is interrupted. Something chimes or jars with an instinctive aesthetic for regularity, and we are inclined to look for causes or explanations. Whatever stands out from the crowd, or breaks from the line, invites us to wonder why. Some argue, plausibly, that we evolved to see a single cause even where there is none, on the basis that it is better to be safe than sorry, better to identify that pattern in the trees as a tiger, better to run—far better—than to assume that what we see is a chance effect of scattered light and shifting leaves in the breeze, creating an illusion of stripes. But this habit, ingrained, defiant, especially if indulged by a snap decision, makes us wretched natural statisticians. Most often the pattern *will be* a chance effect, but we will struggle to believe it. "Where's the tiger?" we say. "No tiger," says the statistician, just chance, the impostor, up to her callous old mischief. In more considered moments, now we have moved on from evolution in the jungle, we should remember our experience of chance, and check the instinct born in moments of haste.

One reason for the punishing cost of medical research is that new drugs have to be trialed in ways that seek to rule out the effect of chance on whether or not they work. You might wonder how hard that can be: administer a drug to a patient and wait to see if the patient gets better, surely, is all there is to it. The joke that a bad cold treated with the latest drugs can be cured in seven days, but left to run its own course may linger for a whole week, shows us what is wrong with that. Patients improve—or not—for many reasons, at different rates. Say we put them in two groups, give one group the drug and another a placebo, and observe that the drugged group improves more. Was it the drug, or chance? Ideally we'll see a big difference between two large groups. But if the difference is small, or few people take part, it's a problem. Like rice grains or cancer cases, some patients in a medical trial will produce results that look meaningful, but are in fact quite unrelated to the influence everyone suspects—the new drug, the phone mast—and everything to do with chance.

Statistics as a discipline has made most of its progress only in the last 200 years. Perhaps the reason it took so long to get started, when science and mathematics had already achieved so much, is that it is a dry challenge to instinct. Particularly in relation to patterns, chance, or coincidence, statistics can feel counterintuitive when it frustrates a yearning for meaning.

"The tiger that isn't" makes a good standard for numbers that seem to say something important but might be random, and there are imaginary tigers in all walks of life. Our question could be this: Is the tiger real? Or are we merely seeing stripes? Is this a pattern of numbers that tells us something, or a purely chance effect that bears only unsettling similarity to something real?

Like illness, events cluster, too. In 2005 three passenger airliners came down in the space of a few weeks, leading to speculation of some systemic problem—"What is causing our planes to fall?" To repeat, chance does not mean without cause—there was a cause for each of those crashes—just separate causes. What chance can do is to explain why the causes came together at the same time, why, in effect, they clustered.

Does this prove that every cluster, cancer or otherwise, is chance alone? Of course not, but we have to rule out that explanation before fastening on another. People in suits seen on the news advising disbelieving residents that their fears are unfounded might be part of a conspiracy against the public interest—it is conceivable—but let's also allow that they speak from acquaintance with what chance is capable of, and have worked honestly to tell stripes from the tiger. The difference often comes down to nothing more than size. One case of a real cluster, in High Wycombe, England, of a rare type of nasal cancer with a genuinely local cause, eventually was attributed to the inhalation of sawdust in the furniture industry, and had prevalence 500 times more than expected. Downwind of the Chernobyl nuclear power station there are a large number of thyroid cancers, far more than even chance might cause.

Whenever we see patterns or clusters in numbers, whenever they seem to have order, we're quick to offer explanations. But the explanation most easily overlooked is that there is no explanation; it was just chance. That doesn't mean everything that happens is due to chance, of course. But in numbers, we need to be every bit as alert for phantom tigers as for real ones. There is more chance about than many of us think.

UP AND DOWN

4

A MAN AND HIS DOG

Brace yourself for a radical fact, a fact surprising to much political debate, capable of wrecking the most vaunted claims of government. Ready? Numbers go up and down.

That's it. No one has to do anything special to cause this. No policy is necessary. In the ordinary course of life, things happen. But they do not often happen with absolute regularity or in the same quantities. Measure (almost) anything you like and on some days there are more, some days fewer. The numbers rise and fall. They just do.

Of course, you knew this. You might think those in authority know it, too. But do they? "We did that!" they holler, whenever the numbers move in a favorable way, apparently oblivious to the possibility that the numbers would have gone up or down anyway.

Mistaking chance ups and downs for the results of new

policies can have dire consequences: it means failure to understand what really works, spending money on things that don't, ignoring what does.

To guard against the problem, think of a man walking uphill with a dog on a long, extendable leash. You can't see the man, it's dark, but the dog's collar is fluorescent, so you watch as it zips up and down, stops and switches. How do you know for sure if the man is heading up, down, or sideways? How do you know how fast? When the dog changes direction, has the man changed, too? Whenever you see numbers go up and down, always ask: Is this the man, or the dog?

Hit the heights in sports or business in America and you might make the cover of those illustrious magazines *BusinessWeek* or *Sports Illustrated*. That is, unless you are also superstitious, when you might do anything to be overlooked.

Sudden shyness in people who bathe in celebrity demands an explanation, and so there is. These magazines . . . (here put on your eye patch and best piratical voice) . . . be cursed. If you believe in such things—and plenty do—then any who dares appear on the cover invites fickle fortune to turn her wheel with a vengeance.

It is true that a surprising number of people or businesses featured this way really do tend to find success suddenly more elusive: firms with a Midas touch lose it, the brightest athletic stars are eclipsed. It is known, in the now famous case of *Sports Illustrated*, as the SI Cover Jinx.

The magazine itself reported on this in 2002: "Millions of superstitious readers—and many athletes—believe that an appearance on *Sports Illustrated*'s cover is the kiss of death," it said, finding 913 jinxes in 2,456 issues "a demonstrable misfortune or decline in performance following a cover appearance roughly 37.2 percent of the time." Eddie Matthews was a Major League baseball player who graced the first cover in 1954. A week later, he injured a hand and missed seven games. There was the skier who broke her neck, numerous winning streaks ended by humiliation to lesser rivals, and many other examples of abrupt bad luck.

Spooky? Preferring statistics to magic, we don't think so. Our skepticism rests on a simple, subversive principle: things go up and down.

Seeking explanation for change, it is easy to overinterpret ordinary ups and downs and attribute them to some special cause—such as a jinx. No magic is required, simply the knowledge that a spectacular season of home runs is seldom beaten next year, or that the skier who takes the most risks might be closest to victory . . . but also to disaster. If you have been "up" sufficient to make the cover of a celebrated magazine, it could be that you are at your peak and, from the peak, there is only down. When the dog reaches the end of the leash, it often runs back. And that, as we say, is it. The jinx is in all probability due to what statisticians call regression to the mean. When things have been out of the ordinary lately, the next move is most likely back to something more average or typical; after a run of very good luck, chance might be ready for revenge, whether you appear on the cover or not.

Similarly, when at the bottom, the tendency is often the other way. "It's a low, low, low rate world" said the cover of *BusinessWeek* about the cost of borrowing in March 2007, shortly before rates rose sharply, though not quite as comically as the 1979 cover predicting "The Death of Equities," when the Dow Jones market index was at 800. In January 2008 it stood at about 12,500. Reporting on this tendency for numbers to go into reverse, *The Economist* magazine confessed its own prediction in the late 1990s that oil would drop to $5 per barrel. In 2008 it topped $140.

Examples like these make it tempting to assume that whatever a business magazine recommends constitutes a sure guide to doing the opposite, being based, as these stories often are, on the ups or downs of the recent past rather than on that more reliable tendency to fluctuation. Alas, there are no sure guides in business, but one study in 2007—"Are Cover Stories Effective Contrarian Indicators?"— by academics at the University of Richmond published in the *Financial Analysts Journal*, looked at headlines from featured stories in *BusinessWeek*, *Fortune*, and *Forbes* magazines from a twenty-year period. It concluded that "positive stories generally indicate the end of superior performance and negative news generally indicates the end of poor performance."

Exposing the cover jinx makes an entertaining but exaggerated point. Not everything that goes up and down is due to chance. In truth, there may be underlying reasons that have nothing to do with chance variation (or magic). But in asking: "Why did it happen?" chance is the potential explanation most often ignored.

And it is not only entertaining. Imagine a forty-year diet of whipped cream and two packs of cigarettes daily. Your arteries quite likely would be as well lined as your furry slippers. A heart attack looms.

Heart attacks are often like that: a long time coming, with causes that tend to be lifelong, the culprits partly genetic, partly dietary, frequently including smoking, stress and lack of exercise—all well-known suspects—and damage that is cumulative. It peaks the day of a good lunch or a tight game on TV, when the quantity of oxygen reaching the heart muscles finally becomes insufficient and . . .

Except, that is, if you live in the small town of Helena, Montana. Here, according to a now famous study, a large proportion of heart attacks has been explained not by people's own habits alone, but partly by those of others. It is breathing cigarette smoke in the atmosphere—passive smoking—that is often to blame. Or so it is said.

The study in Helena was astonishing. It found a fall in heart attacks at the local hospital of 40 percent during a six-month ban on smoking in public places, compared with the same six months previously. When the ban ended, the number of heart attacks went up again. The research was published in the *British Medical Journal*.

We had better declare that neither of us is a pro- nor antismoking lobbyist or campaigner. Neither of us is a smoker, nor wants to breathe other people's smoke, but that preference can't disguise a problem with the Helena study: it overlooks the fact that numbers also go up and down, by chance.

Can chance really explain a 40 percent drop that bounces straight back up? As with clusters, people fidget for meaning when numbers rise and fall. What made them rise; what made them fall? It is right to ask the question, but we need to be prepared for a disappointing answer. Numbers can go up and down for no more reason than occurs when chance varies the height of the waves rolling into shore. Sometimes waves crash, sometimes they ripple, now larger, now smaller. This succession of irregular peaks and troughs, swelling and sinking, fussy and dramatic, we'll call chance. Like chance it has its causes, but so intricately tangled we can almost never unravel them. Imagine those waves like numbers that are sometimes high and sometimes low. It is the easiest mistake in the world to observe a big wave, or a high number, and assume that things are on the up. But we all know that a big wave can arrive on a falling tide.

This is no insight. Everyone knows it from life. The suggestion here is simply that what is known in life is used in numbers. The tendency among some to feel that numbers are alien to experience, incomprehensible, inaccessible, a language they will never speak, should be put aside. In truth, they speak it already.

Numbers often work on principles that we all use with easy familiarity, all the time. And so we can easily see that a momentarily high number can, just like a big wave, be part of a falling trend. And yet taking a measurement from a single peak and assuming it marks a rising tide, or assuming that a trough marks a falling one, though a simple error, is one that in politics really happens. It is assumed that the variation in the numbers tells us all we need to know, that it happened for a critical reason, caused by an action we took or a policy we implemented, when really it is just chance that made the numbers momentarily dip or rise. Is this obvious? It is hard to believe that politics sometimes needs lessons from the rest of us in the view from the beach, but it is true. For to know if the explanation is accurate, what matters is not the size of the wave at all, but rather the tide, or maybe even sea level. The wave catches the eye, but look elsewhere before being sure of capturing a real or sustained change, not a chance one. Otherwise you will fail to distinguish cause and effect from noise.

But to repeat the acid question, can this explain the Helena study? The chart below shows deaths from heart attacks in Lewis and Clark County, Montana. About 85 percent of the county population lives in Helena. These are deaths from heart attacks, not heart attack admissions, but represent the best data we can find covering a sufficiently long period, *before* the smoking ban took effect, in June 2002. The chart also appeared in the *British Medical Journal*, submitted by Brad Rodu, a professor of pathology at the University of Alabama medical school.

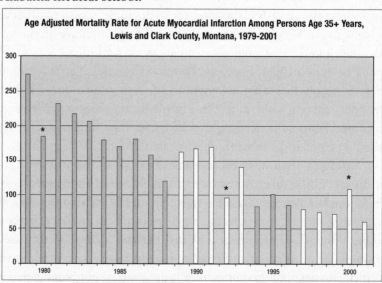

Age Adjusted Mortality Rate for Acute Myocardial Infarction Among Persons Age 35+ Years, Lewis and Clark County, Montana, 1979-2001

Two points stand out: the downward trend (the tide), and that the numbers go up and down around this trend (the waves). And when, by chance, they have gone down sharply in one year, the next move tends to be up. The asterisks mark years in which there was variation of around 30 or 40 percent.

In St. Peter's hospital, Helena, the number of heart attack patients admitted during the study period was tiny: six or seven a month beforehand, an average of four a month during. It is worth adding that most of the fall occurred in the first half of the ban, and that the numbers rose again in the second half. The study was right to be cautious in concluding that laws to enforce smoke-free workplaces and public places "*may be* associated with an effect on morbidity from heart disease" [emphasis ours].

Did the ban on smoking in public places in Helena reduce heart attacks by 40 percent? We doubt it. This looks more like the frisky dog around the man than the man himself. The statistical intelligence of studies on the effect of such bans around the world generally has been poor. A widely reported study in Scotland, based on a sample of hospitals, asserted that a similar ban there reduced heart attack admissions by 17 percent, a huge change. That is, until official data for all admissions showed the true extent of the fall to be about 8 percent, and this against a recent downward trend before the ban of 5 or 6 percent, and a fall seven years earlier of about 11 percent. The fact that there have been several other initiatives designed to reduce heart attacks in Scotland in the last few years makes it hard to say for sure what, if any, effect the ban had. Though we would not be at all surprised if it had some—substantially smaller—effect than that claimed by researchers, and we'd expect that to become clearer with time.

Time would be a good test of whether the fall was a chance, short-term up-and-down, or a real change, but the ban in Helena ended after six months. In California, a ban—together with steep increases in tobacco taxes and an antismoking campaign—is reported to have coincided with a fall in deaths from heart attack about 5 or 6 percent greater than elsewhere in the United States, but this was the cumulative total over a period of nine years. Compare that with a claim of a 40 percent fall in heart attacks in six months from a ban alone, based on a few handfuls of cases.

The inevitability of variation in numbers regularly sinks understanding, as people race to overinterpret what are often chance results. A contentious and none-too-obvious example is the roadside speed camera. In the UK there's been a fierce national debate about the rapid proliferation of speed cameras. The United States, now adopting them in increasing numbers at state level, could learn from the UK's mistakes, where they have become deadly devices, if not to traffic then certainly to conversation. Oppose them and you are assumed to crave toddlers at play slaughtered; you seek to scream unimpeded through residential areas pushing 90 mph. Support them, and it is owing to your control-freak loathing for others' freedom and pig-ignorance of the facts about a flawed system. The antis are caricatured as don't-give-a-damn evaders of civic responsibility, the pros of wanting the motorist fined off the road or filling the cops' coffers. According to the *Daily Telegraph* newspaper (August 10, 2006), "a survey found that 16 percent of people support the illegal destruction of speed cameras by vigilante gangs."

BY LUKE TRAYNOR—*LIVERPOOL DAILY POST*

Merseyside's speed cameras are picking up motorists driving at up to 134 mph, figures revealed yesterday. In one incident, a driver was caught racing at 134 mph in a 50 mph zone, nearly three times over the limit, on the M62 near the Rocket public house. Over a six-month period 116 motorists were caught going at or above 70 mph in a 30 mph zone.

Three cheers for speed cameras? Then again . . .

BY PHILIP CARDY—*SUN*

Nicked for doing 406 mph.

Driver Peter O'Flynn was stunned to receive a speeding notice claiming a roadside camera had zapped him—at an astonishing 406 mph. The sales manager, who was driving a Peugeot 406 at the time, said: "I rarely speed and it's safe to say I'll contest this."

With indignant voices on either side, sane discussion keeps its head down, but why is there even argument? Judgment ought to be easy, based the numbers: are there more accidents with cameras or without? The answer seems clear: on the whole, accidents decline when cameras are installed, sometimes dramatically.

DEPARTMENT FOR TRANSPORT PRESS RELEASE,
FEBRUARY 11, 2003

Deaths and serious injuries fell by 35% on roads where speed cameras have been in operation, Transport Secretary Alistair Darling announced today. The findings come from an independent report of the two-year pilot scheme where eight areas were allowed to re-invest some of the money from speeding fines into the installation of more cameras and increased camera use.

Transport Secretary Alistair Darling said: "The report clearly shows speed cameras are working. Speeds are down and so are deaths and injuries . . . This means that more lives can be saved and more injuries avoided. It is quite clear that speeding is dangerous and causes too much suffering. I hope this reinforces the message that speed cameras are there to stop people speeding and make the roads safer. If you don't speed, you won't get a ticket."

This 35 percent cut in deaths and serious injuries equated, the department said, to 280 people, and since then the number of cameras has grown. Case closed? Not quite. First, there are exceptions: at some sites the number of accidents went up after speed cameras arrived. But a minority of perverse results will not win the argument. We need to know what happens on balance: on balance there was still a large fall.

The second reservation risks sounding complicated, but is in principle easy: it is the need to distinguish between waves and tides, or the dog and the man (which for some years the Department for Transport preferred not to do). As with waves, the rate at which things happen in the world goes up and down, though without telling us much about the underlying trends: there are more this time, fewer next, rising one week, falling another, simply by chance. People especially do not behave with perfect regularity. They do not sort themselves into an even flow, do not get up, go to work, eat, go out, catch the bus, crash, either at the same time or at even intervals. The numbers doing any of these things at any one time goes up and down.

We all know this. Accident statistics at any one site will quite likely go up and down from time to time, because that is what they do—a bad crash one month, nothing the next. The freakish result would be if there were exactly the same numbers of accidents in any one place over every twelve-month period. Luck, or bad luck, often determines the statistics, as the wave rises and falls.

The argument bites once again with the realization that, all else being equal, if the numbers have been higher than usual lately, the next move is more likely to be down than up. After a large wave often comes a smaller one, the dog goes one way then the other. Two bad crashes last month on an unexceptional stretch of road and, unless there is some underlying problem, we would be surprised if things did not settle down.

Applied to road accidents, the principle has this effect: put a speed camera on a site where the figures have just gone up, at the crest of a big

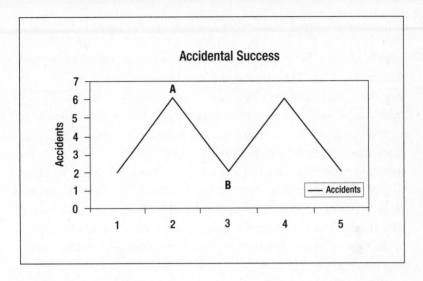

Accidental Success

wave (point A in the above chart)—and this does tend to be when the cameras arrive, in the belief that these sites have been newly identified as a problem—and the next move in the accident statistics is likely to be down—*whether the speed camera is there or not* (point B). The peak turns to a trough, the statistics fall, and presto, the government claims success.

A small experiment shows how pure chance can produce what looks like evidence in favor of speed cameras, using a simple roll of a die. Before describing it, we had better state our view lest we are labeled extremists of one sort or another. On the statistical evidence, most speed cameras probably do cut accidents, some probably do not, depending where they are. So there is probably a benefit, but the size of that benefit has been greatly exaggerated by some—including the government. And this takes no account of the effect that withdrawing police patrols in favor of using roadside cameras has on accidents caused by other driving offenses—drunk driving for example—an effect extremely difficult to estimate. (See also Chapter 7 on risk.)

So, to the experiment. Find a group of people—a typical class size will do but it can be smaller—and ask each person to be a stretch of road, as we did with a gaggle of journalist colleagues. Take your pick: Route 42 in Cleveland, Highway 1764 in Texas, any ordinary stretch of road. Next, everyone rolls a die twice and adds the two throws together. This represents the number of accidents on each stretch of road. And, just by chance, some stretches produce higher numbers of accidents than others, as they do in life, when brakes or a driver's attention, for example, just happen, by chance, to fail at that moment. In our group of about twenty volunteer journalists we then targeted what appeared to be the accident black spots, by identifying all those with a score of 10, 11, or 12. To these we gave a photograph of a speed camera, and asked them to roll again, twice. The result? The speed camera photos are instantly effective. None of the high scorers equaled their previous score once they had a photograph of a speed camera.

It might be objected that these were not the real thing, only photographs, and so the experiment proves nothing. But the point is that at some genuine roadside accident sites placing a photograph—or even pebble—on the pavement would have been as effective as a camera, simply because the rise and fall in the numbers of accidents has been due to chance, just like a six on the roll of a die. Since we have put the cameras (or the photographs of a camera, or the pebble) in a place where the number has just, by chance, been high, it is quite likely, whatever we do, by chance to go down. It looks as if we had something to do with it. In fact, we just caught the wave and took the credit.

It is another case of regression to the mean—when a number has reached a peak or trough lately, the next move is likely to be back toward the average—and was initially discounted by British Department for Transport researchers, despite our challenge to their figures. Then, after claiming for two years a large number of lives saved by cameras, estimates of the benefit were cut sharply in 2006 when, for the first time, some attempt was made to calculate how much of this was due to regression to the mean, and chance was finally given her due.

Even after revision the DfT figures were still unsatisfactory, being the result of a dubious mixture of definitions and some sloppy arithmetic. Figures in the ministerial press release had to be corrected after we challenged them. But in their more modest claims they probably were at least a little closer to the truth.

Of the full fall in the number of people in fatal or serious collisions at speed camera sites, about 60 percent was now attributed to regression to the mean, and about another 18 percent to what are called "trend" effects, namely that the number of accidents was falling everywhere, including where there were no cameras, as a result, for example, of improved road layout and car safety. This leaves, according to the department's revised figures, about 20 percent of the apparent benefit of speed cameras genuinely attributable to them, though even this figure is contested.

To observe the tide and not the wave of accidents at speed-camera sites, there are two options: (a) wait until we have seen plenty of waves come and go, which in the case of speed cameras is now believed to be for about five years; (b) do not begin measurements only from the peaks, but do plenty of them, and choose points of measurement at random. More simply, we can remember that there is a lot of chance about.

Public policy worldwide has a truly shocking history of ignorance about whether the benefits it claims in a precious few oft-quoted examples have occurred entirely as a result of chance or other fortuitous association, or if the policy genuinely makes the differences claimed. At the Home Office, a 2006 report on the evidence for the effectiveness of policy to tackle reoffending found that not one policy reached the desired standard of proof of efficacy, because so many had failed to rule out the possible effects of chance in numbers when counting the rise or fall in offenses committed by people on various schemes.

That doesn't mean that nothing works. It means we do not know if it works, because we did not take account of the fact that numbers go up and down—and particularly that they tend to go down when they have been high, or go up when they have been low. A senior adviser at the Home Office who knows what a sneaky devil chance can be, and how easily the numbers can mislead, says to ministers who ask what actually works to prevent reoffending: "I've no idea."

Can this really be how government proceeds? Without much by way of statistical rigor and cursed with a blind spot for the effects of luck and chance? All too often, it is.

That tendency to ignore the need for statistical verification is only now beginning to change, with the slow and often grudging acceptance that we need more than a plausible anecdote (a single wave) before instituting a new policy for reoffenders, for teaching methods, for health care, or any other state function. Politicians are among the most recalcitrant, sometimes pleading that the genuine pressure of time, expense, and public expectation makes impossible the ideally random-controlled trials that would be able to identify real stripes from fake, sometimes apparently not much caring or understanding, but, one way or another, often resting their policies on little more than luck and a good story, becoming as a result the willing or unwilling suckers of chance.

A politician with a taste for a calculated gamble is a disappointingly welcoming way in for chance to do its dirty work. A more shocking example comes with the two innocent doctors who felt confronted—on nothing more than the turn of a roulette wheel—with an insinuation of murder.

It has happened. Two general practitioners, summoned to a meeting, sat tense and anxious, trying to explain the unusual death rates of their patients. The meeting took place in the shadow of Dr. Harold Shipman—a Yorkshire GP recently convicted of murdering probably well in excess of 200 people. An inquiry into the Shipman case, led by Dame Janet Smith, was trying to establish whether GP patient mortality could be monitored in order to spot any other GPs who gave cause for concern. These two doctors had been identified from a sample of about 1,000 as among a dozen whose patients were dying at rates as high as, or higher than, Shipman's.

An initial statistical analysis had failed to find an innocent explanation. What other remained, except for quality of care—for which
read the hand of the two GPs? This was where Dr. Mohammed
Mohammed came in. He was one of those asked to investigate.
"This was not a trivial meeting," he told us. One can imagine. It was
deeply uncomfortable, but necessary, he added, in order to explore
any plausible alternatives. He was not about to throw at anyone a
casual accusation of murder, but they had all read the newspapers. A
softly spoken, conscientious man, determined to proceed by scientific method, he wanted to know before coming to a conclusion if
there were any other testable hypotheses.

The statistical analysis that brought the GPs to this point had
sought to sift from the data all variation due to chance or any unusual characteristics of their patients: the ordinary rise and fall in the
numbers dying that would be expected anywhere, together perhaps
with an unusually elderly population in the area, or some other characteristic in the case-mix that made deaths more frequent, were obvious explanations. But these were examined and found wanting.
Now it was the GPs' turn.

And the reason they offered, essentially, was chance; chance
missed even by a statistical attempt to spot and control for it. No,
their patients were not significantly older than others, but the GPs
did have, by chance, an unusually high number of nursing homes on
their patch. And while age alone is often a good proxy for frailty or
increased illness, a nursing home is much better. People here are
most likely to be among the most frail or in poor health.

We know without doubt that there was nothing accidental about
Shipman, but the variation of death rates, even variation on a par
with mass murder, can have a perfectly innocent explanation, as
chance ruffles the smoothness of events, people, and places, changing
a little here and there, bringing a million and one factors into play,
suggesting vivid stripes without a tiger to be seen.

To see the problem of telling one from another, think of your signature. It varies, but not too much, and what variation there is worries no one—it is normal and innocent. Try with the other hand, though, and the degree of variation probably jumps to the point where you hope that on a check the bank would reject it. The task in detecting a suspect pattern is to know where the normal chance variation stops, and what is known as "special cause" variation begins. How varied can a signature be and still be legitimate? How much variation in mortality rates can there be before we assume murder? The answer is that innocent or chance variation can exceed even the deliberate effect of Britain's most prolific serial killer.

The GPs' explanation turned out to be entirely consistent with the data, once examined closely, mortality rates at the nursing homes being perfectly in line with what would be expected—for nursing homes. They were cleared completely, but hardly happy. If the sample of GPs that identified these two was anything to go by, we would expect to find about 500 across the UK with mortality rates similarly inviting investigation. That is an awful lot of chance at work, but then, chance is like that: busy—and cunning, and with about 40,000 GPs to choose from in the UK, some certainly would be unlucky even if none was murderous.

Dame Janet Smith said in her final report from the Shipman inquiry: "I recognize, of course, that a system of routine monitoring of mortality rates would not, on its own, provide any guarantee that patients would be protected against a homicidal doctor. I also agree with those who have emphasized the importance, if a system of routine monitoring is to be introduced, of ensuring that PCTs (Primary Care Trusts), the medical profession and the public are not lulled into a false sense of security, whereby they believe the system of monitoring will afford adequate protection for patients."

Aware of these difficulties, she recommended GP monitoring all the same, but justified as much by the value of the insight it might offer into what works for patient care—a view shared by Dr. Mohammed—as for detecting or deterring murder. Can we really not tell chance death from deliberate killing? With immense care, and where the numbers are fairly clear-cut, as in the Shipman case, we can, just about. But he was able to avoid detection for so long because the difference was not instantly obvious then and, with all our painfully learned statistical wisdom since, the chilling conclusion is that we are only a little better at spotting it now, because chance will always fog the picture. It was in February 2000, shortly after Shipman's conviction, that the then Secretary of State for Health, Alan Milburn, announced to Parliament that the department would work with the Office of National Statistics "to find new and better ways of monitoring deaths of GPs' patients." This promise was still unfulfilled seven years later. That is a measure of the problem—and testimony to the power of chance.

When statisticians express caution, this is often why: chance is a dogged adversary. And yet it can be overcome. Trials for a new polio vaccine in the United States and Canada in the 1950s had to meet persistence with persistence. At that time, among 1,000 people, the chances were that none would have polio. It always was quite rare. So let's say a vaccine comes along and it is given to 1,000 people. How do we know if it worked? Chance probably would have spared them anyway. How can we tell if it was the vaccine or chance that did it?

The answer is that you need an awful lot of people, so that chance is by and large ironed out by sheer weight of numbers. In the trials for the Salk polio vaccine, nearly 2 million children were observed in two types of study. The number who for one reason or another did not receive the vaccine, either because they were in one of the various control groups, or simply refused, was 1,388,785. Of these, 609 were subsequently diagnosed with polio, a rate of about one case in every 2,280 children.

Among the nearly half million who were vaccinated, the rate was about 1 case in nearly 6,000 children. The difference was big enough, among so many people, that the research teams could be confident they had outwitted chance's ability to imitate a real effect. Though even here they looked closely to make sure there were no chance differences between the groups that might have accounted for different rates of infection. Getting the better of chance can be done, with care and determination, and often involves simply having more stamina or patience than she does.

One more example, which we will look at in more detail in a later chapter. School exam results go up and down from year to year. They move so much, in fact, that their relative performances—how one compares with another—are shuffled substantially each year. But is it the school's teaching standards that are thrashing up and down? Or is the difference due to the ups and downs in pupils' ability as measured by exams from one year to the next? It seems mostly the latter, and that is as you might expect: what principally seems to determine a school's exam results is the nature of its intake. In the UK, in fact, the results for a school in any one year are so subject to the luck of the intake that for between two thirds and three quarters of all schools, the noise of chance is a roar, and we are unable to hear the whisper above it of a real influence or a special cause; we are unable to say with any confidence whether there is any difference whatsoever that is made by the performance of the schools themselves. Chance so complicates the measurement that for the majority of schools the measurement is, some say, worthless. We would not go nearly that far. But we would agree that data is regularly published that misleadingly implies these waves are a reflection of the educational quality of the school itself, and report year-to-year changes in performance as if they were clear indicators of progress.

A rising tide or just a wave? The man or the dog? Stripes or a real tiger? We can be vigilant, we need to be vigilant, but we will be fooled again. The least we can do is determine not to make it easy for chance to outwit us. That task is begun by knowing what chance is capable of, a task made easier if we slow the instinctive rush to judgment and beware the tiger that isn't.

AVERAGES 5

THE WHITE RAINBOW

verages play two tricks: first, they put life's lumps and bumps through the blender. It might be bedlam out there but, once averaged, the world turns smooth. The average wage, average house prices, average life expectancy, the average crime rate, as well as less obvious averages like the rate of inflation . . . there are ups and downs mixed into them all. Averages take the whole mess of human experience, some up, some down, some here, some there, some almost off the graph, and grind the data into a single number. They flatten hills and raise hollows to tell you the height of the land—as if it were flat.

But it is not flat. Forget the variety behind every average and you risk trouble, like the man who drowned in a river that rose, he heard, on average only to his knees. So trick one brings a problem: it stifles imagination about an awkward truth—that the world is a hodgepodge of uneven variety.

Trick two is that averages pass for typical when they may be odd. They stand for what's ordinary, but can be warped by what's exceptional. They look like everyman but can easily be no one. They sound like they're in the middle, but may be nowhere near.

The way to see through an average is to picture the variety it blends together. An image might help make that thought vivid: "White, on average," is what we would see by mixing the light from a rainbow, then sharing it equally. But it is a wretched summary of the view. It bleeds from the original all that matters—the magical assortment of colors. Whenever you see an average, think: "white rainbow," and imagine the vibrancy it conceals.

In his final State of the Union address, in January 2008, President George W. Bush argued for the continuation of tax cuts introduced in 2001, on behalf of a great American institution: the average taxpayer.

"Unless the Congress acts," he said, "most of the tax relief we have delivered over the past seven years will be taken away." He added that 116 million American taxpayers would see their taxes rise by an average of $1,800.

Which was true, approximately. The original tax-cut legislation had an expiry date. Unless renewed, the cut would end and taxes would go up. The Tax Policy Center, an independent think tank, calculated that the average tax increase would be $1,713, close enough to the president's figure of $1,800. And so the typical American citizen could be forgiven for thinking: the president means me.

Actually, no he didn't. He might like you to think he did, but he probably didn't. About 80 percent of taxpayers would pay less than $1,800, most of them a lot less. That is, more than 90 million of the 116 million taxpayers would *not* see their taxes rise by this much. To many, that feels intuitively impossible. How can so many people, so many more than half, be below average? Isn't what's average the same as what's typical?

That confusion served the president well. Being opposed to tax rises, he wanted this one to appear big to as many people as possible, so he gave the impression that the typical experience would be an $1,800 tax hike when, in truth, only one in five would pay that much.

How did he do it? He used the blender. Into the mix, he poured all taxpayers, from street sweepers to the richest yacht-collecting hedge-fund manager. Nothing wrong with that, you might say. But so rich are the richest that you can dilute them with millions of middle and low incomes and the resulting blend is still, well, rich.

Even though everyone is in it, this average is not typical. Think of the joke about four guys in a bar when Bill Gates walks in. They whoop and holler.

"Why the fuss?" asks Bill, until one of the four calms down to answer:

"Don't you know what you've just done to our average income?"

Or think about the fact that almost everyone has more than the average number of feet. This is because a few people have just one foot, or none at all, and so the tiny influence of a tiny minority is nevertheless powerful enough to shift the average to something a bit less than two. To have the typical complement of two feet is therefore to have more than average. Never neglect what goes into an average, and why the influence of a single factor, one part of the mix, might move the whole average somewhere surprising and potentially misleading.

The chart divides taxpayers into groups, 20 percent of them in each, and shows the increased tax liability for each group. The tax rise for the middle 20 percent would average not $1,800, but about $800. The bottom fifth would pay an average extra $41. We can see, in fact, that four out of the five groups, accounting for about 80 percent of

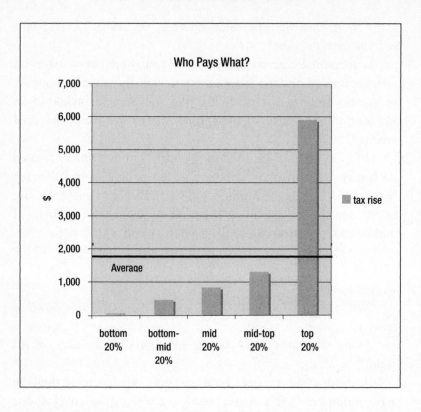

taxpayers, would pay less than the average. We can also see what lifts the average—the tall column representing the income of the richest 20 percent.

If we looked at this last group in more detail, we would find that the top 0.1 percent would pay an extra $323,621 each. These are income centipedes, millipedes even, who are so rich, paying so much tax, that they move the average far, far more than other individuals.

This is not an argument about the merit of tax cuts in general or about how they were implemented by President Bush. It is about what happens when they are presented using an average. The president did the same when the tax cut was introduced, selling it on the basis of the average benefit for all taxpaying Americans, when the benefit was massively skewed to those who paid most tax. Once again, you might agree with that emphasis, but it is a bias the average conceals. Averages are like that: in trying to tell us about an entire group, they can obscure what matters about its parts. This applies not only to economics, where averages are often cited, but to almost any description of typical experience: to take a surprising example—pregnancy.

Victoria Lacey was pregnant and overdue in early September 2005. Her due date had been August 26. Two weeks late and pregnancy becoming exasperating, she began each day hoping: "Let this be the one." As the long hours passed, she became resigned to another. Was something wrong? "Why can't your body produce a baby on the date it's supposed to?" she asked herself.

But which date is that? Doctors give expectant mothers an estimated date since, naturally, they can never be certain, and that estimate is based on the average length of a pregnancy. But how long is an average pregnancy? The answer, unhelpfully, is that the official average pregnancy is probably shorter than it ought to be.

There were 669,601 live births recorded in the UK (2006) and 4,138,349 in the United States (2005, latest data); is it possible that every due date was misleading? Some will have been right by chance, simply because pregnancies vary in duration, but will they have been right as often as they could? The impression of an imprecise science is confirmed when we learn that the practice in France is to give a latest date, not a due date, some ten days later. Victoria gave birth to baby Sasha, safely and without inducement, about two weeks overdue, on September 10, 2005.

Due dates in the United States and UK are initially calculated by counting 280 days from the first day of the last menstrual period. Doctors settled on this number in part because it seemed about right, but also under the influence of a Dutch professor of medicine named Herman Boerhaave ("So loudly celebrated, and so universally lamented through the whole learned world" —Samuel Johnson).

Boerhaave wrote nearly 300 years ago that the duration of pregnancy was well known from a number of studies. Those studies have not survived, though their conclusion has. It remains well known up to the present day, consolidated by influential teachers and achieving consistency in medical textbooks by about the middle of the twentieth century. Some are also familiar with Naegele's Rule, based on the writings of Franz Naegele in 1812, who said that pregnancy lasted ten lunar months from the last menstrual period, also giving us 280 days. Nearly everyone in the United States and UK still agrees that 280 days is correct: it is the average.

But averages can deceive. A drunk sways down the street like a pendulum from one pavement to the other, positioned on average in the center of the road between the two white lines as the traffic whistles safely past, just. On average, he stays alive. In fact, he walks under a bus.

Averages put variation out of mind, but somewhere in the average depth of the river there might be a fatally deep variation; somewhere in the distribution of the drunk's positions on the road there was a point of collision that the average obscures. The enormous value of the average in contracting an unwieldy bulk of information to make it manageable is the reason why it can be so misleading.

So what is going on in the rich variety of experiences of pregnancy? In particular, what happens at the edges? Two facts about pregnancy suggest that the simple average will be misleading. First, some mothers give birth prematurely. Second, almost no one is allowed to go more than two weeks beyond the due date before being induced. Premature births pull the average down; late ones would push it up, but we physically intervene to stop babies being more than two weeks late. The effect of this imbalance—we count very early births but prevent the very late ones—is to produce a lower average than if nature were left to its own devices. That is not a plea to allow pregnancies to continue indefinitely, just to offer a glimpse inside the calculation.

We might argue in any case that very premature births ought not to be any part of the calculation of what is most likely. Most births are not significantly premature, so if a doctor said to a woman: "The date I give you has been nudged down a few days to take account of the fact that some births—though probably not yours—will be premature," she might rightly answer, "I don't want my date adjusted for something that probably won't happen to me, I want the most likely date for what probably will." The average is created in part by current medical practice—the numbers we've got arise partly through medical intervention—and the argument in favor of 280 days becomes circular: this is what we do because it is based on the average, and this becomes the average partly because of what we do.

Mixing in the results from the edges produces a duration that is less likely to be accurate than it could be. In the largest recent study, of more than 400,000 women in Sweden, most had not yet given birth by 280 days—the majority of pregnancies lasted longer. By about 282 days, half of babies had been born (the median), but the single most common delivery date (the mode), and thus arguably the most likely for any individual in this distribution, was 283 days.

If most women have not had their baby until they are at least two days overdue, and women are more likely to be three days overdue than anything else, it invites an obvious question: are they really overdue?

You are forgiven for finding it muddled. Yet above all this stands the unarguable fact, confirmed by all recent studies, that the duration at which more women have their baby than any other is 283 days. This type of average—the most common or popular result—is known as the mode, and it's not clear why in this case it isn't preferred. In fact, that latest study from Sweden found that even the simple arithmetical average (the mean)—the one that adds in all the premature births—was not in fact 280 days, but 281.

None of this would matter much (one, two, or three days is sometimes frustrating, but in a normal pregnancy most likely to be medically neither here nor there), were it not that these numbers form part of the calculation of when to induce birth artificially. Induction is often offered to women in this country, sometimes with encouragement, seven days after the due date. It raises significantly the likelihood of Caesarean section—which has risks of its own—and can be, for some, a deeply disappointing end to pregnancy.

When obstetricians answer that induced birth has better outcomes for women than leaving them a little longer, they might also tell you that one of the ways they measure that "better outcome" is by asking women if they feel better having sorted out the problem of being overdue. If you tell someone there's a problem, they may well thank you for solving it. If they knew the problem was based on a miscalculation, they might feel otherwise.

Averages bundle everything together. That is what makes them useful, and sometimes deceptive. Knowing this, it is simple to avoid the worst pitfalls. All you need do is remember to ask: "What interesting flavors might be lost in that blend?" If they tell me the rainbow is white on average: "What colors are blanked out?"

So we need to go carefully when offered an average as a summary of disparate things. It would help if journalists, politicians, and others took care to avoid using them as a way of saying "just about in the middle" (unless this is an average called the median), avoided using them to stand for what is "ordinary," "normal," or "reasonable," avoided using them even to mean "most people," unless they are sure that is what they meant. They may be none of those things.

The "middle" is a slippery place altogether. Middle America, much like Middle Britain, is a phrase beloved of politicians and journalists. Middle America is primarily a place, but bundled up with social and moral values and the economic plight said to be typical of decent

American citizens, who are neither rich, nor poor, but hardworking, probably families and, you know, sort of, customarily, vaguely, in the middle. It is a target constituency for candidates of all political parties, given extra appeal by being blended with an idea of the middle class, a group-membership now claimed by the vast majority. A survey by the American Tax Foundation found that four out of five Americans label themselves "middle class." Just 2 percent call themselves "upper class."

In sum, the "middle" has become impossibly crowded. Politicians particularly like it that way, so that any proposal said to benefit even some small part of the middle-class/middle-America/average-citizen can leave as near as possible the entire population feeling warm and adored, at the heart of a candidate's concern. Vague? You bet. By this standard, we're all in the middle now.

Whatever use these middles have as a description of values, they are hopeless as a classification of where people stand within the rainbow variety of incomes or economic circumstances.

In the United States, the household truly in the middle of the income distribution (known as the median) has income of about $48,000 before taxes (in 2006). (If two people in a household have incomes, the statistics add them together, and call it household income.) The middle 20 percent of households are between about $38,000 and about $60,000. The middle 60 percent—to stretch the membership—are between about $20,000 and $97,000.

Should all these households belong in the same economic and social bracket? How much in common does one family have with another when its income is five times higher? Democratic presidential contender John Edwards said in 2007 that $200,000 was middle class (though $250,0000 wasn't). Just before the Pennsylvania primary in 2008, Hillary Clinton was reported in *The New York Times* to have promised not to raise taxes on "middle-class Americans, people making less than $250,000 a year," even though $250,000 a year would take a household into the top 3 percent, more than five times the income of the household truly in the middle and twelve times that of some households in the middle 60 percent. When concepts become this elastic, they belong to comic-book superheroes.

The Tax Foundation has pointed out that people often compare themselves to others in their own community, and that a household in a rich neighborhood—Fairfax County, Virginia, to take the highest income county in the United States—would be plumb in the local middle on about $100,000 a year. By contrast, a household in St. Landry Parish, Louisiana, would be bang in its middle on about $24,000.

Our sense of the middle, of what's average, depends on who is around us. We slip into thinking that what is true here is true everywhere. Our sense of where we fit is also heavily weighted, in the case of incomes, by the feeling that our own are nothing special, even when, by comparison, they soar. Both influences mean we are poor judges of the American income rainbow, and have a generally lousy sense of what is typical. Much media comment about the middle, economically speaking, has no idea where it is.

In the 2005 UK General Election, Charles Kennedy, then leader of the Liberal Democrat Party, found himself ducking the notion of the average income like a sword of truth, swung in defense of the ordinary citizen.

His party had said we should be rid of a local tax known as the council tax, replacing it with a local income tax. Some—the richest—would pay more, most would pay less. Then came the inevitable question: how much did you have to earn before local income tax raised your bill? As Charles Kennedy stumbled, at a loss for the accurate figure, the journalists at the press conference sniffed their story. The answer, when it came in a scribbled note from an adviser, gave the Liberal Democrats their roughest ride of the campaign: A little over £40,000 ($80,000) per household.

The problem was that this figure turned out to be about twice average earnings. To take the example every news outlet seized, a

firefighter on average earnings living with a teacher on average earnings would pay more under the Liberal Democrats' proposed new
system.

The campaign trail suddenly was hazardous for Charles Kennedy,
the questions and coverage indignant: what was fair about a system
that hit people on average earnings? If it is going to be more expensive
for the average, how can you possibly say most people will be better
off? So, the Lib Dems are hitting Middle England? And so on.

There might be other good reasons for opposing the policy, but
there is also such a hodgepodge of misconceptions about the idea
that lay at the root of the criticism—the average earner—that it is
hard to know where to start.

As in America, as with pregnancy, or feet, so with income: the
average is not in the middle, and more people are on one side than
the other. A UK household with two individuals each earning
average individual earnings is not in the middle of the household
income distribution; it is, in fact, in the top quarter; it is, if you are
wise to the rainbow distribution of incomes and not cocooned by
the even larger salary of a national newspaper commentator, to be
relatively rich. This is for two reasons: first, because in this case the
average is pulled well beyond the middle by a relatively small number
of enormously high incomes; second, because it is relatively unusual
for both members of a couple to have average earnings. In most cases
where one half of a couple has average earnings, the other will earn
significantly less. It might surprise us to find that a teacher and a
firefighter living together are relatively rich, but only if we do not
know how they compare with everyone else, don't know where they
are in the distribution, and have ignored the colors of the income
rainbow.

The next chart shows the distribution of income in the UK for
childless couples—two people living together, their incomes combined. Half have net incomes (after tax and benefits) of less than
£18,800 (about $38,000) (marked as the median), but the average for

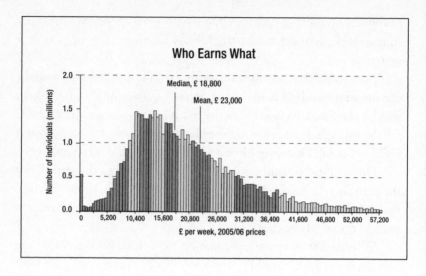

the group is around £23,000 (about $46,000), pulled up by the relatively small numbers of very high incomes. That is, most are at least 18 percent below average. The highest incomes are far too high to fit on the chart, which would need to stretch yards to the right of the edge of the page to accommodate them. The most common income is around £14,000 (about $28,000), roughly 40 percent below average. More people in this category have incomes at this level than any other. It was intriguing to tell these numbers to colleagues in the BBC, on what for them was a modest income of about £50,000 ($100,000) a year, and watch their jaws hit the floor. Some knew roughly the average; few if any knew what was typical. Economically speaking, the "middle Britain" of media imagination is a mythical country, conjured up to give a respectable home to their prejudices, regardless of the economic facts, or what is truly typical.

A Dutch economist, Jan Pen, famously imagined a procession of the world's population where people were as tall as they were rich, everyone's height proportional to their wealth (note wealth, not income). A person of average wealth would be of average height. The procession starts with the poorest (and shortest) person first and ends, one hour later, with the richest (and tallest).

Not until twenty minutes into the procession do we see anyone at all. So far, they've had either negative net worth (owing more than they own) or no wealth at all, and so have no height. It's a full thirty minutes before we begin to see dwarfs about six inches tall.

And the dwarfs keep coming. It is not until forty-eight minutes have passed that we see the first person of average height and average wealth, when more than three quarters of the world's population has already gone by.

What delays the average so long after the majority have passed? The answer lies in the effect of those who come next. "In the last few minutes," wrote Pen, "giants loom up . . . a lawyer, not exceptionally successful, eighteen feet tall." As the hour approaches, the very last people in the procession are so tall we can't see their heads. Last of all, said Pen (at a time before the fully formed fortunes of Bill Gates and Warren Buffett), we see John Paul Getty. His height is breathtaking, perhaps ten miles, perhaps twice as much.

One millionaire can shift the average more than many hundreds of poor people, one billionaire a thousand times more. They have this effect to the extent that 80 percent of the world's population has less than average.

In everyday speech, "average" is a word meaning low or disdained. With incomes, the average is high. The colloquial use, being blunt, thoughtless, and bordering on a term of abuse, distorts the statistical one, which might, according to the distribution, be high, or low, or in the middle, or altogether irrelevant. It is worth knowing which. If only one thought survives about averages, let it be that they are not necessarily anywhere near the middle, nor representative of what's typical, and that places often called "the middle" by politicians or the media may be far removed. These ideas have been lazily hitched together for too long. It is time for a divorce. We must have a little background before we know the real relationship.

. . .

The writer and paleontologist Stephen Jay Gould had special reason to find out. Diagnosed with abdominal mesothelioma in 1982, a rare and highly dangerous cancer, at the height of a brilliant career and with two young children, he quickly learned that it was incurable and that the median survival time after discovery was eight months. He gulped, he wrote later, and sat stunned for fifteen minutes.

Then, he says, he started thinking. He describes the ensuing story—a statistical one—as profoundly nurturing and life-giving.

The median is another variety of average: it means the point in a distribution where half are above and half below, what Gould and others call a central tendency. In this case it meant that half of all people diagnosed with abdominal mesothelioma were dead within eight months.

A number like that hits brutally hard. But it is worth recalling that other numerical characteristic that applies particularly to averages of all kinds, namely, that they are not hard and precise, do not capture some Platonic essence, are not, as Stephen Jay Gould understood to his advantage, immutable entities. Rather, averages are the classic abstraction, and the true reality is in the distribution, or as Gould says: "in our actual world of shadings, variation and continua."

In short, wrote Gould, "we view means and medians as the hard 'realities' and the variation that permits their calculation as a set of transient and imperfect measurements." In truth, so far as there are any hard realities at all, they are in the variation, in the vibrant individual colors of the rainbow, not in the abstracted average that would declare the rainbow white.

Once again, to make sense of this average, we must ask what is in the distribution. The first part of the answer is straightforward: half will live no more than eight months, that much is true, but half will also live longer, and Gould, being diagnosed early, reckoned his chances of being in the latter half were good.

The second part of the answer is that the eight-month median tells us nothing about the maximum for the half who last longer than eight months. Is eight months halfway, even for the luckiest? Or is the upper limit, unlike the lower one, unconstrained? In fact, the distribution had a tail that stretched out to the right of the graph for several years beyond the median, being what statisticians call right-skewed. If you lived longer than eight months, there was no knowing how much longer you might live beyond that.

Understanding the distribution behind the average allowed Gould to breathe a long sigh of relief: "I didn't have to stop and immediately follow Isaiah's injunction to set thine house in order for thou shalt die, and not live," he wrote. "I would have time to think, to plan, and to fight." He lived, not eight months, but another twenty years, and when he did die, in 2002, it was of an unrelated cancer.

The same skewed distribution lies behind the average life span for people in Swaziland, the lowest in the world according to the 2007 CIA Factbook and United Nations social indicators, at thirty-two years for men and thirty-three for women. It is a frighteningly low figure. But thirty-two is not a common age of death—most who make it that far survive longer—it is simply that in order to calculate the average their fates are bundled together with the shocking number who die in infancy. The average is pulled down by Swaziland's high rates of infant mortality. And so the figure for average life expectancy stands like a signpost between two destinations, pointing at neither. It fails to convey either of what we might call the two most typical life spans, but is an unsatisfactory blend of something truly awful and something more hopeful, a statistical compromise that describes very few. A better way of understanding life expectancy in Swaziland would be to say that it is either only a few years, or that it is not very far from our own. Prospects there are largely polarized; the average yokes the poles together.

Should averages simply be avoided? No, because for all their hazards, sometimes we want a figure to speak for a group as a whole, and averages can be revelatory. Making them useful is mainly a question of working out which group we are interested in. What we most often want to know when we find ourselves reaching for an average about some social, economic, or political fact of life, is what is true for most people or most often, what is typical, what, to some extent, is normal for a particular group. Some of the many reasons an average might not tell us any of these things are found above, to do with the awkward failure of life to behave with regularity. But say we are pig-headed and still want to know. Then there is nothing for it: with all those caveats about what might be going on in the misty reaches of the distribution, we still need to find words for a summary. And sometimes that summary of the typical will be what we most want to know about the success or failure of a policy.

A good current example in the UK is hospital waiting lists. How long people wait for treatment is a thorny subject in UK health care. So the government set a target—currently that no one should wait more than six months (twenty-six weeks) for an operation. This is about to change to the much more demanding target of a maximum of thirteen weeks from GP referral to treatment, continuing the deliberate focus on one end of the distribution, the long-waiting end. Once there was a long tail to the waiting list, and you could wait years for treatment, but it is now an abrupt cutoff with a maximum waiting time for all. It reminds us how certain parts of the distribution rather than the whole can acquire political importance.

As a result of this policy, the government became fond of saying that waiting times were coming down (Department of Health press release, Wednesday, June 7, 2006: "Waiting times for operations . . . shorter than ever"), and it was true that the longest waits had been reduced dramatically. But the longest waits, though a real problem, were a small proportion of the total number of waits. Not very many were ever kept waiting more than two years compared with the millions who were seen within a few months. So this was a case where we might also want to know, in order to say with authority what is happening generally to waiting times, what is happening to everyone else, not only to the long waits, but also to the more typical waits, those who always were inside the maximum. The government had taken one part of the distribution and talked about it as if it spoke for the whole. How do we find a more satisfactory measure?

The best answer in this case is to ask what happens to a patient whose wait is somewhere in the middle of all waits, the point where half of all patients wait less and half more, and which is an average called, as in the Stephen Gould case, the median. These figures were decidedly mixed, and in some cases startling. Even so, this example will not take the median for every patient in the country, but will identify the median for various large groups.

Before we look at them, we will also do one other thing to the calculation of waiting times. We will strip out a practice that seems to have increased significantly in recent years of resetting the waiting-time clock for what are sometimes opportunistic reasons (see Chapters 7 and 6 on counting and targets). This is permitted but easily abused—the hospital rings you to offer an appointment with one day's notice, you can't go, they reset your waiting time. Once you strip out this behavior to compare like with like waits, five years ago with now, and look at the typical patient, not just the very longest waits, the effects are illuminating.

The UK health service is divided into areas called primary care trusts (PCTs). In one that we looked at (in late 2006), for example, waiting times for trauma and orthopedics had gone up for the typical patient from 42 days to 102 days. In another there was an increase from 57 days to 141 days. A third saw a rise from 63 days to 127 days.

Waiting times for ear, nose, and throat (ENT) told a similar story, where the typical patient in about 60 percent of primary care trusts was waiting as long or longer than five years before. In general surgery, figures for more than half of all PCTs showed that the typical patient was waiting longer than five years earlier. In a number of cases, though not a majority in the areas we looked at, even the waits for the 75th percentile (how long before 75 percent of patients were treated) had gone up.

Then, in March 2008, the UK media was shocked at the revelation that median waiting times had gone up across the board, from forty-one days to forty-nine days, and one of the central political claims of the last ten years dissolved.

The key point of our analysis is that it makes little sense to say waiting times are going one way or another unless you say "for whom?" and identify the group that interests you. The figures are not moving the same way for everyone, and in some large categories of treatment, though not for all, most patients, as typified by the median, are waiting as long as or longer than five years ago.

It is important to know what has happened to the longest waits and it is quite reasonable to say that shortening them is a success for those patients, but it makes no sense to say that these alone tell you whether "waiting times have come down." They do not. For this, there is no option but to use some kind of average, and the most appropriate average is the median.

Incidentally, we found that most hospitals we asked could not tell us what was happening to the typical (median) patient, and the initial data came from another source—the Dr. Foster organization, what is now known as a health informatics company—which had access to the raw hospital episode statistics and the in-house capacity to crunch the data.

Always ask about an average: which group are we really interested in? Maybe when we ask about the average income, we don't want to know about the stratospheric earners, we want to know about what's more typical. And maybe there are other strange colors in other averages that we don't want in the mix, as well as some that we do. What matters is that you know what's in and what's out, and that you have achieved the mix you want. Averages are an abstraction, a useful one, but an abstraction all the same. If we look at them without knowing what it is that we have abstracted from, we will be misled. It is an average, but an average of what? Remember the vibrancy and variety of real life. Remember the rainbow.

PERFORMANCE 6

THE WHOLE ELEPHANT

Pick one number to tell your life story, a single measure to sum up your life's worth. What's it to be? Height? Weight? How about salary? For some, that would do nicely. Most would feel cheapened. How about your personal tally of wives or husbands? Or maybe longevity captures it? Up to a point, perhaps, until someone asks what you did with all those years. Whatever the measurement, if one part is taken for the whole, it is open to ridicule. One number, because it implies one definition, is almost never enough.

Social and political life is as rich and subtle as our own, and every bit as resistant to caricature in a few digits. If you want to summarize like this, you have to accept the violence it does to complexity.

This is why targets and performance measurements often struggle. They typically pick one aspect of performance, a measurable aspect, then set a quantified target for its improvement, but put everything else aside. They try to glimpse a protean whole through the keyhole of that single number. The strategy for seeing through them is much like that for the average: think about what they do not measure, as well as what they do; think about what else lies beyond the keyhole, not only what you can see through it.

An Indian fable, *The Blind Men and the Elephant*, is best known to Western readers in a version by the American poet John Godfrey Saxe (1816–1887):

> *It was six men of Indostan*
> *To learning much inclined,*
> *Who went to see the Elephant*
> *(Though all of them were blind),*
> *That each by observation*
> *Might satisfy his mind*

But their conclusions depend entirely on which part of the elephant they touch, so they decide, separately, that the elephant is like a wall (its side), a snake (trunk), spear (tusk), tree (leg), fan (ear), or rope (tail).

> *And so these men of Indostan*
> *Disputed loud and long,*
> *Each in his own opinion*
> *Exceeding stiff and strong,*
> *Though each was partly in the right,*
> *And all were in the wrong!*

A whole elephant is a devil to summarize. A single measure of a single facet leaves almost everything useful unsaid, and our six men still in the dark. In life, there are many elephants.

But the problem is worse than that. In health and education (two of the biggest), it is not only that one part doesn't adequately represent the whole. There is also the tendency for the parts we do not measure to do odd things when our backs are turned: while measuring the legs, the trunk starts to misbehave. And so there have been occasions in U.S. health care when the chosen measure was the number of patients surviving an operation, with the result that some surgeons avoided hard cases (who wants patients who spoil the numbers?). At least if they died without reaching the operating table, they didn't risk dying on it. So part of the elephant, unseen or at least unmeasured, was left to rot, even as the measured part told us the elephant was healthy.

There is huge interest in performance measurement, for obvious reasons. In the U.S. economy $1 out of every $7 spent is spent on health care, more than any other country, yet political arguments rage about whether it is well spent, particularly when an estimated 40 million people lack basic health-care coverage. That is not to claim the system doesn't work; it is simply an observation that Americans themselves argue fiercely about whether it is efficient. For both privatized and state-run systems, efficiency is an objective all share and, on both sides of the Atlantic, similar, performance-related incentives in search of efficiency are now in vogue, sometimes bringing real improvement, but also often bringing the same elephantine problems.

So to be partly in the right and partly wrong, like the six blind men, is often the best that targets and performance indicators can do, by showing only part of the picture. The ideal would portray the whole elephant, but it is no slight on numbers to say that they cannot. If this means observing the people who come off the operating table, but not those who were kept away, we can't complain at being duped: after all, it is we who put on the blinkers.

Targets, and their near allies performance indicators, face just such a dilemma. One measurement has to stand as the acid test, one number to account for a wide diversity of objectives and standards, while the rest of the elephant . . . out of sight, undefined, away from scrutiny, unseen through the keyhole, who cares about the rest?

This suggests that the best strategy with any single-number summary of performance is to be clear not only what it measures, but what it does not, and to perceive how narrow that definition is. So that when a good health service is said to mean, as it has been in the UK, short waiting times, and so waiting times are what are measured in the hope of concentrating people's minds on making them shorter, someone might stop to ask: "and is the treatment at the end *any good*?" Sadly, no one has worked out how to measure the quality of health care with any subtlety, yet. We are left instead with a proxy, the measurement that can be done rather than the one that ideally would be done, even though it might divert us from what we really want to know—a poor shadow that might look good when quality is bad, or vice versa—right in part, and also wrong.

In a famous cartoon mocking old-style Soviet central planning and target setting, the caption told of celebrations after a year of record nail production by the worker heroes of the Soviet economy, and the picture showed the entire year's glorious output: a single gigantic nail; big, but big enough for what? Measurements can be a con unless they are squared up to their purpose, but it is a con repeatedly fallen for.

According to Gwyn Bevan, professor of management science at the London School of Economics, and Christopher Hood, professor of government and fellow of All Souls College Oxford, the current faith in performance measurement rests on two "heroic" assumptions.

The first is the elephant problem: the parts chosen must robustly represent the whole, a characteristic they call "synecdoche," a figure of speech. For example, we speak of hired hands, when we mean, of course, the whole worker. The second heroic assumption is that the design of the target will be "game proof."

The difficulties with the second follow hard on the first. Because one part seldom adequately speaks for all, it gives license to all kinds of shenanigans in the parts that are untargeted, out of sight, and often out of mind. Hence the name "gaming," or, as it is often described, "hitting the target but missing the point." So if your life did happen to be judged solely on your income, with no questions asked and the rest of your life out of sight beyond the keyhole, "gaming" would be too gentle a word for what you might get up to at the bank with shotgun and stocking mask.

Thus in the United States over the years, there has been a long list of attempts to measure (and therefore improve) performance, often with financial incentives, that have somehow also returned a kick in the teeth. The following are just a few examples from studies published in various medical journals.

In New York State it was found that reporting of cardiac performance had led to reluctance to treat sicker patients and "upcoding" of comorbidities (exaggerating the seriousness of the patient's condition). This made it look as if the surgery was harder, so that more patients might be expected to die, meaning that performance looked impressive when they didn't.

More than 90 percent of America's health plans measure performance using a system called HEDIS (Healthcare Effectiveness Data and Information Set). This consists of seventy-one measures of care. In 2002 it was found that, rather than improve, some poor performers simply had ceased publishing their bad results.

In the 1990s, the prospective payment system was introduced for Medicare. This paid a standard rate for well-defined medical conditions in what were known as diagnosis-related groups (DRGs). The system effectively set a single target price for each treatment for all health-care providers, thus encouraging them to bring down their costs, or so it was thought.

The effect was satirized in an article in the *British Medical Journal* in 2003 that feigned to offer advice. "The prospective payment system in the United States ... has created a golden opportunity to maximize profits without extra work. When classifying your patient's illness, always 'upcode' into the highest treatment category possible. For example, never dismiss a greenstick fracture as a simple fracture—inspect the X-ray for tiny shards of bone. That way you can upgrade your patient's break from a simple to a compound fracture and claim more money from the insurance company. 'DRG creep' is a well-recognized means of boosting hospital income by obtaining more reimbursement than would otherwise be due."

The article added that a national survey of U.S. doctors showed 39 percent admitted to using tactics—including exaggerating symptoms, changing billing diagnoses, or reporting signs or symptoms that patients did not have—to secure additional services felt to be clinically necessary. The scope for such behavior has been reduced in the years since, but not eliminated.

When we asked people to e-mail *More or Less* with examples of their personal experience of "gaming" in the British health service, we had little idea of the extent, or variety.

Here is an example. A consultant was urgently asked to come to the emergency room. She hurried down to find a patient about to breach the four-hour target for bed waits. She also found another patient, who seemed in more pressing need, and challenged the order of priority. But no, they wanted her to leave that one: "He's already breached."

Here is another: "I work in a specialist unit. We're always getting people sent to us by Accident and Emergency (A & E) because, once they are referred, the patient is dealt with as far as they are concerned and has met their target. Often, the patients could have been seen and treated perfectly well in A & E. But if they were sent to us, they might be farther away from home and have to wait several days to be treated because they are not, and frankly never were, a priority for a specialist unit."

And another: "I used to work for a health authority where part of my job was to find reasons to reset the clock to zero on people's waiting times. We were allowed to do this if someone turned down the offer of an appointment. But with the waiting-time targets, we began working much harder to find opportunities to do it, so that our waiting times would look shorter."

Some people respond to a system of performance measurement by genuinely improving their performance, but others react by expending their energy arguing with it, some ignore it, some practice gaming, finding ways, both ingenious and crude, to appear to be doing what's expected but only by some sleight of hand, and some respond with downright lies about their results.

Bevan and Hood identify four types. There are the "saints," who may or may not share the organization's goals, but their service ethos is so high that they voluntarily confess their shortcomings. Then there are the "honest triers" who won't incriminate themselves, but do at least get on with the job without resorting to subterfuge. The third group they call the "reactive gamers," who might even broadly agree with the goals but, if they see a low-down shortcut, take it. Finally, there are what Bevan and Hood call "rational maniacs," who would spit on the whole system, but do what they can to conceal their behavior, shamelessly manipulating the data.

Given these and other complications, good numbers could be telling us any of four things. 1: All is well, performance is improving, and the numbers capture what's going on. 2: The numbers capture what's going on in the parts we're measuring but not the weird things now happening elsewhere. 3: Performance as measured seems fine, but it's not what it seems because of gaming. 4: The numbers are baloney, lies. Unfortunately, we often can't tell which.

Bevan and Hood have catalogued examples of targets in the health service that aspired to something good in itself but pulled the rug on something else. The target was hit but the point missed and damage done elsewhere.

In 2003, a parliamentary committee found that the waiting-time

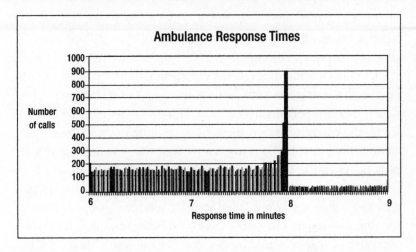

target for new ophthalmology outpatient appointments had been achieved at one hospital by cancelling or delaying follow-up appointments. As a result, over two years at least twenty-five patients were judged to have gone blind.

In 2001 the government said all ambulances should reach a life-threatening emergency (category A) within eight minutes, and there was a sudden massive improvement in the number that did, or seemed to. But what is a "life-threatening emergency"? The proportion of emergency calls logged as category A varied fivefold across trusts, ranging from fewer than 10 percent to more than 50 percent. It then turned out that some ambulance services were doctoring their response times—lying, to put it plainly. They cheated on the numbers, but it was also numbers that found them out (together with a sharp-eyed researcher named Richard Hamblin), when it was discovered that there was a strangely dense concentration of responses recorded just inside eight minutes, causing a sudden spike in the graph, and almost no responses just above eight minutes. This was quite unlike the more rounded curve of response times that one would expect to see, and not a pattern that seemed credible. The chart shows the pattern that led to suspicion that something was

amiss. There was even some evidence that more urgent cases sometimes were made to wait behind less urgent (but still category A)
cases, to meet the targets.

On UK waiting times in the emergency room, where there has
been a sharp improvement in response to a target, but ample evidence of misreporting, Bevan and Hood concluded: "We do not
know the extent to which these were genuine or offset by gaming
that resulted in reductions in performance that was not captured by
targets." Gwyn Bevan, who is, incidentally, a supporter of targets in
principle and has even helped devise some, told us that when a manager worked honestly but failed to reach a target, then saw another
gaming the system, hitting the target and being rewarded, the strong
incentive next time would be to game like the rest of them: bad behavior would drive out good.

And on *More or Less* we found our own evidence of gaming in
the United Kingdom when we observed the increasing numbers
coming to the emergency room, but found a much greater increase
in the proportion of them who were formally admitted to the hospital. There were two reasons. One was a target to treat or admit all
patients within four hours. But it turned out that many of the extra
admissions were for a very short period. This made it appear as if
everyone was dealt with inside four hours. But some were admitted
only to be kept waiting for another half hour, then treated and dis-
charged. The second incentive was that each admission was worth
£500 to the hospital under a national tariff, even if, again, only for
half an hour. Hard-pressed doctors in a busy department, who knew
they could meet their targets *and* pick up £500 for the hospital simply
by moving the patient down the corridor a few yards and calling this
an admission, often did so. But did it represent an improvement in
performance?

Then there was a new contract for family doctors, which began
in 2004, and rewards them for, among other things, ensuring that
they give patients at least ten minutes of their time (a good thing, if

the patients need it). This gives an incentive to spin out some consultations that could safely be kept short, "asking after Aunty Beryl and her cat," as one newspaper put it. The creditable aim was to make sure people received proper attention. The only measurement available was time, and so a subtle judgment was summarized with a single figure, which became a target, which created incentives, and which led to suspicions of dumb behavior.

The key point from this mountain of evidence is that when we use numbers to try to summarize performance, all this, and more, will be going on in the background, changing even as we attempt to measure it, even changing *because* we attempt to measure it. Can numbers capture the outcome reliably? In essence, there is an unending struggle between the simplicity of the summary and the complexity (and duplicity) of human behavior.

In health data there's yet another twist to that complexity. A paper by Rodney A. Hayward, from the University of Michigan, in 2007, points out that performance measures for health care are often agreed after high-stakes political arguments in which anyone with an interest in an illness advocates idealized standards of treatment. They want, naturally, more and more resources for their cause and press for care that may be only marginally beneficial. They fight for standards on behalf of the most needy cases.

But this entirely understandable advocacy for idealized standards of treatment for say, diabetes, takes no account of the demands for treatment of, say, Alzheimer's. So performance measurement against a set of such standards can make it appear that everyone is failing.

Hayward comments: "It sounds terrible when we hear that 50 percent of recommended care is not received, but much of the care recommended by subspecialty groups is of modest or unproven value, and mandating adherence to these recommendations is not necessarily in the best interests of patients or society . . . Simplistic all or nothing performance measures can mislead providers into prioritizing low-value care . . ."

In other words, performance measurement would work best if it took into account the needs of the whole elephant, alongside the costs, something which advocacy groups, or the process of piecemeal standard setting, almost by definition, don't do.

Health care involves choices. If we are to make them well, we need to recognize that the ideal standard of care for one condition might mean sacrificing standards for another. It is no good simply to say that we want the best of everything, unless we are prepared to bankrupt ourselves for it. In practice, we need to assess the claims of every treatment against the claims of every other. These claims will vary from patient to patient. How do we choose between them if we have already laid down rules in advance for how each one should be treated? The more we specify, the less we can choose.

There is no simple answer to this. We do not advocate renouncing all rules or performance measures any more than we think that everything should be specified. But we do have to understand the risks, of which there are broadly two: either that people cheat, or that they do exactly what you ask, and then, when this turns out to be at the expense of everything else, you sort of wish they hadn't.

If we trusted them to get on with the job as well as it could be done, none of this would arise, but we don't, sometimes with reason. So we intervene and try to change behavior, hoping to steer it with numbers.

Though the most conspicuous use of targets and performance indicators has been in health care, they are spreading. In the UK, they have arguably become the most trusted tool for managing the public sector. Trusted by the government, that is. For while the numbers have often shown improvement, not everyone believes them, and ministers struggle with the public perception that increased spending on public services has been poor value for money. In short, despite the hope that measurements and targets would ensure efficiency, many suspect that they have actually disguised waste.

In the United States, too, there are signs of a growing belief that problems can be solved by insisting that the numbers improve. But which numbers? The elephant problem is not confined to health.

Ask what education should consist of, for example, and you hear a hundred different answers. Look how the quality of a U.S. high school is measured and you find one: multiple-choice examination success in math and English.

Since the late 1980s, education policy in the United States has tended toward setting targets for students. Children are no longer compared one with another in the same class to see who's best, but measured against a fixed academic standard for all, set at state level, which they meet or fail. This policy trend has been most fully realized by No Child Left Behind (NCLB), the crux of the Bush education policy, but enjoying bipartisan support. It was approved in 2001 and was due for renewal in 2008—the presidential election permitting. Every school is also expected to improve its results to meet Adequate Yearly Progress (AYP) benchmarks.

NCLB seems to have brought improvements, but with some costs, and is controversial. One acid dispute has been over the high-stakes motivation. Failure to achieve the standards set under NCLB has severe consequences: staff can be fired, students can be held back a year or even denied a graduation diploma, states can be forced to bus students to different schools and spend money on individual tuition.

It may be true that math and English are the most important subjects in secondary education, but so much incentive for one measure means that if this is all we look at through the keyhole, this is possibly all we will get. The risk is that what we count, will count, and what we don't, won't. That means, according to some critics, less art, science, physical education, or IT. It might mean that some students will be neglected: the strugglers who haven't much hope and the able who can be depended on to pass. Borderline students, on the other hand, will be worth investment of time and effort. This has been colorfully described in the United States as educational triage, when so-called bubble kids close to proficiency receive more attention at the expense of others, although one study of 300,000 students in a western state reported little evidence of this in math test scores. But the fear persists.

There is evidence that some schools have manipulated graduation rates, another public measure of performance, by taking them as a percentage of the number of students enrolled in later years, rather than the number enrolled at the beginning of high school. As the *New York Times* reported in March 2008: "When it comes to high school graduation rates, Mississippi keeps two sets of books. One team of statisticians working at the state education headquarters here recently calculated the official graduation rate at a respectable 87 percent, which Mississippi reported to Washington. But in another office piled with computer printouts, a second team of number crunchers came up with a different rate: a more sobering 63 percent . . .

"As a result, researchers say, federal figures obscure a dropout epidemic so severe that only about 70 percent of the one million American students who start ninth grade each year graduate four years later."

There are reports that some schools encourage poorly performing pupils to take an easier and different test, so they are not even entered for the test on which the school is measured. Some schools apparently have boosted their pass rate by counting successful retakes a year later, but not counting the initial failure.

Other examples from the UK include Britain's proud record of having the safest roads in Europe. We measure this by the number of accidents. But there are both elephant and gaming problems with our road-safety statistics. The first comes about because we define road safety by what happens on the roads. But one way to cut accidents is to make roads so fast and treacherous to pedestrians—as two-lane highways and the like—that no pedestrian goes near them. Risk aversion is not the same as safety. In some ways, the roads might be more dangerous even though casualties are falling.

But are casualties falling? Over the very long term they are, unquestionably. More recent evidence for adult casualties is less clear. The government has targeted road accidents, telling police forces they would be judged by their success in reducing the number killed and seriously injured on the roads. By 2010, it says, there should be 40 percent fewer accidents overall than in the baseline period of 1994 to 1998. As the target was introduced, the numbers began to fall and the government hailed a dramatic success.

Then, in July 2006, the *British Medical Journal* reported an investigation of trends in accident statistics. This said that, according to the police, rates of people killed or seriously injured on the roads fell consistently from 85.9 accidents per 100,000 people in 1996 to 59.4 per 100,000 in 2004.

But the police are not the only source for such statistics and the *BMJ* authors' inspiration was to check them against hospital records. There, recorded rates for traffic injuries were almost unchanged at 90 in 1996 and 91.1 in 2004. The authors concluded that the overall fall seen in police statistics for nonfatal road traffic injuries "probably represents a fall in completeness of reporting of these injuries."

Deaths on the roads, where there is little scope for statistical discretion, have been largely flat in recent years in both police and hospital statistics. But the police have some discretion about how they record an injury, and it is in this category of their statistics that the bulk of the improvement seems to have occurred, an improvement not corroborated by hospital records. So did accidents really go down, or did the police simply respond to the target by filling fewer notebooks?

One last example in a list long enough to suggest a generic problem. In response to concern about recycling rates in Britain lagging behind others in Europe, the government targeted them. Local councils responded with ingenuity and started collecting waste they had never tried to collect before, but that they could easily recycle. They called it, to give the whole enterprise a lick of environmental respectability, green waste. We do not really know what happened to this waste before: some probably was burnt, some thrown on the compost, some no doubt went in the black bin with the other trash. Now a big truck came to collect it instead. Being heavy with water (waste is measured by weight), the vegetation did wonders for the recycling rate. There have even been stories of green waste being sprayed with water to make it heavier. But is that really what people had in mind when they said there should be more recycling?

If all this leads to the conclusion that measurement is futile, then it is a conclusion too far: whether your income is $1 a day or $100 matters enormously and we can measure that pretty well. The key is to know the number's limitations: how much does it capture of what we really want to know? How much of the elephant does it show us? When is it wise to make this number everyone's objective? How will people behave if we do?

This suggestion—that targets and other summary indicators need to be used with humility—implies another; that until they are, we must treat their results with caution.

After years of bruising experience, there are signs of such a change of approach. Britain's Healthcare Commission, responsible for monitoring performance in hospitals across England and Wales, no longer gives the impression of thinking that it can make fine distinctions between the quality of one hospital and another. Instead, it puts more of its energies into two strategies: the first is to set bottom-line standards of practice, not outcomes, across all areas of health care that all are expected to achieve: are they cleaning the place properly, have staff been trained to report adverse incidents, and so on.

The second, more interesting, is spotting the real problems, those whose performance is so out of line that it is unlikely to be a statistical artifact. Not the great bulk of hospitals and procedures that seem broadly all right or better—these are left largely alone—but those with a pattern of results that cause concern. This is surveillance, not performance measurement, or at least not as we have previously understood it. The Healthcare Commission calls its approach risk-based: there is a huge difference between the conceit of performance measurement and the practical importance of spotting potential danger. When they think the numbers suggest a problem—and they generally are restrained in that judgment—they do not necessarily assume anything definitive, or go in with a big stick. So an investigation of "apparently" high rates of mortality at Staffordshire NHS trust, particularly among emergency admissions, was duly cautious with its suspicions. Nigel Ellis, the Commission's head of investigations, said: "An apparently high rate of mortality does not necessarily mean there are problems with safety. It may be there are other factors here such as the way that information about patients is recorded by the trust. Either way it does require us to ask questions, which is why we are carrying out this investigation."

On top of this, they conduct a large number of spot checks, both random and targeted, to try to ensure that the data matches reality. If they find that hospitals have been falsely reporting standards, the hospitals are penalized. This seems to be working. Hospitals that concealed bad performance in the past and were found out seem to turn into saintly confessors in succeeding years.

It is the data that really stands out that has a chance of telling us something useful, whether for better or worse. It is here, in our view, where the numbers offer good clues, if not definitive answers, here that anyone interested in improving performance should start work, never assuming that the numbers have done that work for them.

The Healthcare Commission still does not routinely investigate possible areas of gaming, but there is now such pressure to ensure the integrity of the data that this too might not be far off.

In the past, there has been an incentive not to bother. Both the target setter—usually the government—and target managers want the numbers to look good, and critics allege collusion between them. Some of the early measurements of waiting times in the UK, for example, were a snapshot during a short period that—laughably—was announced well in advance, giving hospitals a nod and a wink to do whatever was necessary in that period, but not others, to score well. It was known that hospitals were diverting resources during the measured period to hit the target, before moving them back again.

It is as well there is room for improvement, since Bevan and Hood say that targets and performance indicators are not easily dispensed with: alternatives like command and control from the center are not much in favor either, nor does a free market in health care escape the demand for performance measurement. If the authorities were serious about the problem of gaming, an interesting approach would be to be a bit vague about which performance measure will be used, so that no one is quite sure how to duck it (will it be first appointments or follow-ups?), introduce some randomness into the monitoring, and there could be more systematic monitoring of the integrity of the numbers.

They conclude: "Corrective action is needed to reduce the risk of the target regime being so undermined by gaming that it degenerates, as happened in the Soviet Union."

We haven't yet reached that point of ridicule, though it might not be far off. The Police Federation, the organization representing rank and file police officers in the UK, has complained (May 2007) that its members are spending increasing time prosecuting trivial crimes and neglecting more important duties simply in order to meet targets for arrest or summary fine. Recent reports include that of a boy in Manchester arrested under firearm laws for being in possession of a plastic pistol, a youth in Kent for throwing a slice of cucumber at another youngster, a man in Cheshire for being "in possession of an egg with intent to throw," a student fined after "insulting a police horse." Perhaps the best (or worst) example is the child who, according to a delegate at a Police Federation conference, went around his neighborhood collecting sponsorship money, then stole it. After a lengthy investigation, the police had to decide whether he had committed one offense or dozens. Since they knew the culprit and all the victims, they said "dozens," as this did wonders for their detection rate. The absurdity of such fatuous activity—all in the name of improving police performance—is that it will give the appearance of an increasing crime rate.

Underlying many of the problems here is the simple fact that measurement is not passive; it often changes the very thing that we are measuring. And many of the measurements we hear every day, if strained too far, may have both caricatured the world and so changed it in ways we never intended. Numbers are pure and true; counting never is. That limitation does not ruin counting by any means, but if you forget it, the world you think you know through numbers will be a neat and tidy illusion.

RISK

7

BRING HOME THE BACON

Numbers have amazing power to put life's anxieties into proportion: Will it be me? What happens if I do? What if I don't? They can't predict the future, but they can do something almost as impressive: tame chaos and turn it into probability. We actually have the ability to measure uncertainty.

Yet this power is squandered through an often-needless misalignment of the way people habitually think, and the way risks and uncertainties are typically reported.

The news says, "Risk up 42 percent," a solitary, abstract number. All you want to know is, "Does that mean me?" There you are, wrestling with your fears and dilemmas, and the best you've got to go on is a percentage, typically going up, and generally no help whatsoever.

Our thoughts about uncertainty are intensely personal, but the public and professional language can be absurdly abstract. No surprise, then, that when numbers are mixed with fear the result is often not the insight it could be, but confusion, and fear out of proportion.

It need not be like this. It is often easy to bring the numbers back into line with personal experience. Why that is not done is sometimes a shameful tale. When it is done, we often find that statements about risk that had appeared authoritative and somehow scientific were telling us nothing useful.

The answer to anxiety about numbers around risk and uncertainty is, like other answers here, simple: be practical and human.

Don't eat bacon. Just don't. That's not a "cut down" or a "limit your intake," it's a "no." This is the advice of the American Institute for Cancer Research (AICR): avoid processed meat. "Avoid" means leave it alone, if at all possible.

The Institute says: "Research on processed meat shows cancer risk starts to increase with any portion." And the Institute is right; this is what the research shows. A massive joint report in 2007 by the AICR and the World Cancer Research Fund found that an extra ounce of bacon a day increased the risk of colorectal cancer by 21 percent. A single sausage is just as dangerous.

You will sense that there is a "but" coming. Let it wait, and savor for a while the authority of the report, even if you can no longer savor a bacon sandwich, in the AICR's own words:

"[The] Expert Report involved thousands of studies and hundreds of experts from around the world. First, a task force established a uniform and scientific method to collect the relevant evidence. Next, independent research teams from universities and research centers around the world collected all relevant literature on 17 different cancers, along with research on causes of obesity, cancer survivors and other reports on chronic diseases. In the final step, an independent panel of 21 world-renowned scientists assessed and evaluated the vast body of evidence."

All this is true. As far as it is possible to discern the effect of one part of your diet, lifestyle, and environmental exposure from all the others, the evidence is not bad, and has been responsibly interpreted. So what's the "but"?

The "but" is that nothing we have said so far gives you the single most essential piece of information, namely, what the risk actually is. We've told you how much it goes up, but not where it started, or where it finished. In the reporting of risk by the media and others, that absurd, witless practice is standard. Size matters to risk: big is often bad, small often isn't—that's the whole point of quantification. You want to know if the food on your plate is of the same order as Russian roulette, crossing the road, or breathing. It makes a difference, obviously, to whether you eat it. But in far too many reports, the risk itself is ignored.

Percentage changes depend entirely on where you start: double a risk of one in a million (risk up 100 percent!) and it becomes two in a million; put an extra bullet in the revolver and the risk of Russian roulette also doubles. But all the newspaper tells you is what the risk has gone up by (100 percent in both cases). By this standard, one risk apparently is no worse than the other.

This—you might think conspicuous—oversight is strangely (and in our view scandalously) typical. "Risk of X double for pregnant women" . . . "Drinking raises risk of Y" . . . "Cell-phone cancer risk up 50 percent." You'll be all too familiar with this type of headline, above reports that often as not ignore the baseline risk.

Let's do it properly. What is the baseline risk of colorectal cancer? There are two ways of describing it—an obscure one and an easy one. Skip this paragraph if you don't fancy obscurity. First the way the AICR report does it. The incidence of colorectal cancer in the United States at the moment is about 45 per 100,000 for men and about 40 per 100,000 for women. That's on page twenty-three. A hundred pages later, we find the 21 percent increase owing to bacon. None of this is intuitively easy to understand or conveniently presented. Media coverage was by and large even worse, usually ignoring the baseline risk altogether.

Fortunately, there is another way. Those who neglect it, whether media, cancer charities, or anyone else, ought to have a good explanation, though we're yet to hear one. And we need endure neither the "eat bacon and die" flavor of the advice from some quarters, bland reassurance from others, nor the mire of percentage increases on rates per 100,000 from others.

Here it is:

About five men in a hundred typically get colorectal cancer in a lifetime. If they all ate an extra couple of slices of bacon every day, about six would.

And that's it.

All the information so mangled or ignored is there in two short sentences which, by counting people, not abstract, relative percentages, are intuitively easy to grasp. You can see for yourself that for 99 men in 100, an extra couple of slices of bacon a day makes no difference to the risk of colorectal cancer, and you can decide whether to take the risk of being the one exception.

"Save our Bacon: Butty Battle!" said the notoriously populist *Sun* newspaper, in a reference to the overfilled traditional English sandwich. But it beat the serious newspapers for intelligible reporting of the risks, being one of very few to make it clear how many *people* could be affected.

It should be easy. And yet . . .

"For every alcoholic drink a woman consumes, her risk of breast cancer rises by 6 percent."

That's pure garbage, by the way, despite its prominence on the BBC's TV news bulletins in November 2002, and it's soon obvious why: if true, every woman who drank regularly—and plenty who liked an occasional tipple—would be a certainty for breast cancer in time for Christmas. A 6 percent increase with every glass soon adds up; about seven bottles of wine over a lifetime would make you a sure thing.

Not much in life is this certain, and breast cancer certainly isn't. Still, ludicrous implausibility didn't stop the claim making the headlines. With luck, viewers were less easily taken in than the journalists who produced the report, since this is a passing piece of innumeracy that should have been easy to spot.

So what was the real meaning of the study so mangled in the news? It was true that research had shown a link between alcohol and breast cancer, but the first thing to do when faced with an increase in risk is to try to quell the fear and concentrate on the numbers. And the first question to ask about this number couldn't be simpler, that old favorite: "How big is it?"

Cancer Research UK, which publicized the research—a perfectly professional piece of work, led by a team at Oxford University—announced the results as follows:

A woman's risk of breast cancer increases by 6 percent for every extra alcoholic drink consumed on a daily basis, the world's largest study of women's smoking and drinking behavior reveals.

It added that for two drinks daily, the risk rises by 12 percent. This was how most other news outlets reported the story. So there was something in this 6 percent after all, but it was 6 percent if you had one drink every day of your adult life, not 6 percent for every single drink, a small change of wording for a not-so-subtle difference of meaning.

This at least is accurate, but also still meaningless. What's missing, yet again, is how big the risk was to start with. Until we know this, being told only by how much it has changed is no help at all.

It is plain to see that the same percentage rise in risk can lead to a very different number at the end depending on the number at the beginning. It is astonishing how often news reports do not tell us the number at the beginning, or at the end, but only the difference.

"Local teenage pregnancies up 50 percent." But is that from 2 last year to 3 this year, or from 2,000 to 3,000? Viewers of the TV quiz show *Who Wants to Be a Millionaire?* know that doubling the money with the next correct answer varies hugely depending how much you have won already. "Contestant doubles her money" tells us nothing.

So why does the news sometimes report risk with only one number, the difference between what a risk was and what it becomes? *"Risk for drinkers up 6 percent!"* Six percent of what? What was it before? What is it now? These reports, whatever their authors think they are saying, are numerical pulp.

How do we make the numbers for breast cancer meaningful? First, once more, the formal way, which you can ignore again if inclined. We need to know the baseline risk—the risk of contracting the disease among women who do not drink. About 9 percent of women will have breast cancer diagnosed by the time they are eighty. Knowing this, we can get an idea of how much more serious it is for drinkers, and because the baseline risk is fairly small, an increase of 6 percent in this risk will still leave it relatively small. Like the slow runner, a 6 percent increase in speed will not make him a contender.

(To do the calculation properly, take the roughly 9 percent risk we started with and work out what 6 percent of that would be: 6 percent of 9 percent is about 0.5 per cent. That is the additional risk of drinking one unit daily, a half of 1 percent, or for two drinks daily about 1 percent. But that is still not as intuitively easy to grasp as it could be. Many people often struggle to understand percentages in any form. In a survey, 1,000 people were asked what "40 percent" meant: (a) one quarter, (b) four out of ten, or (c) every fortieth person. About one third got it wrong. To ask, as we just did: "what is 6 percent of 9 percent?" will surely baffle even more.)

Two days after the study made so many headlines Sarah Boseley, the *Guardian* newspaper's level-headed health editor, wrote an article entitled "Half a Pint of Fear": "How many of us poured a glass of wine or a stiff gin last night without wondering, if only briefly, whether we might be courting breast cancer? . . . There will undoubtedly be women who turned teetotal instantly."

Now why would they do that? Only, as she went on to suggest, because they had been panicked by the 6 percent figure when in truth the difference a drink a day made could have, and should have, been presented in a less alarming light. That is, unless creating alarm was the whole point.

Our interest is not advocacy, but clarity. So let us start again without a practice that makes the risk appear as big as possible. Let's do away with percentages altogether and speak once again, as journalists should, of people.

Here is the simpler way to describe what the reports horribly failed to do, looking at the effect of two drinks a day, rather than one, to keep the numbers round.

"In every 100 women, about 9 will typically get breast cancer in a lifetime. If they all had two extra drinks every day, about 10 would."

Once more, you can quickly see that in a hundred women having two alcoholic drinks every day there would be about one extra case of cancer.

Though 1 woman in 100 is a small proportion, because the American population is large, this would still add up to quite a few cases of breast cancer (if all women increased their drinking by this amount). Our point is not to make light of a frightening illness, nor to suggest that the risk of cancer is safely ignored. It is because cancer is frightening that it is important to make sense of the risks in a way most people can understand. Otherwise we are all at the mercy of news reports that appear like bad neighbors, leaning over the fence as we go about our lives and saying with a sharp intake of breath: "You don't want to be doing that." And maybe you don't. But let us take the decision on the basis of the numbers presented in a way that makes intuitive human sense.

The number of people affected in every hundred is known by statisticians as a natural frequency. It is not far from being a percentage, but is a little less abstract, and that helps. For a start, it is how people normally count, so it feels more intuitively intelligible. It also makes it much harder to talk about relative differences, harder to get into the swamp of talking about one percentage or another. By the way, if you think we exaggerate the difficulty for many people of interpreting percentages, we will see in a moment how even physicians who have been trained in medical statistics make equally shocking and unnecessary mistakes when interpreting percentages for their patients' test results.

Natural frequencies could easily be adopted more widely, but are not, so tempting the conclusion that there is a vested interest both for advocacy groups and journalists in obscurity. When the information could be conveyed in so plain a fashion, why do both often prefer to talk about relative percentage risks without mentioning the absolute risk, all in the most abstract terms? The suspicion must be that this allows the use of "bigger" numbers ("6" percent is big enough perhaps to be a scare, the absolute change "half of 1 percent," or even "1 woman in every 200" is probably less disturbing). Bigger numbers win research grants and sell causes, as well as newspapers.

One standard defense to this is that no one is actually lying. That is true, but we would hope for a higher aspiration from cancer charities or serious newspapers and broadcasters than simply getting away with it. The reluctance to let clarity get in the way of a good story seems particularly true of health risks, but there are, in fact, international guidelines on the use of statistics that warn against the use of unsupported relative risk figures. Cancer Research UK seems either to have been unaware of, or to have ignored, these guidelines in this case in favor of a punchier press release. When we have talked to journalists who have attended formal training courses in the UK, not one has received any guidance about the use of relative risk figures.

In January 2005 the president of the British Radiological Protection Board announced that risks revealed in new medical research into cell phones meant children should avoid them. The resulting headlines were shrill and predictable.

He issued his advice in the light of a paper from the Karolinska Institute in Sweden that suggested long-term use of cell phones was associated with a higher risk of a brain tumor known as an acoustic neuroma. But how big was the risk? The news reports said that cell phones caused it to double.

Once again, almost no one reported the baseline risk, or did the intuitively human thing and counted the number of cases implied by that risk; the one honorable exception we found—all national newspapers and TV news programs included—being a single story on BBC News Online. A doubling of risk sounds serious, and it might be. But as with our two men in a race, it could be that twice as big, just like twice as good, or twice as bad, doesn't add up to much in the end.

With cell phones you could begin with the reassurance that these tumors are not cancerous. They grow, but only sometimes, and often slowly or not at all after reaching a certain size. The one problem they pose if they do keep growing is that they put pressure on surrounding brain tissue, or the auditory nerve, and might need cutting out. You might also note that the research suggested these tumors did not appear until after at least ten years of at least moderate cell phone use. And you might add to all those qualifications the second essential question about risk: how big *was* it? That is, what was the baseline?

When we spoke to Maria Feychting of the Karolinska Institute, one of the original researchers, a couple of days after the story broke, she told us that the baseline risk was 0.001 percent or, expressed as a natural frequency, i.e., people, about 1 in 100,000. This is how many would ordinarily have an acoustic neuroma if they didn't use a cell phone. With ten years regular phone use, the much-reported doubling took this to 0.002 percent, or 2 people in every 100,000 (though it was higher again if you measured by the ear normally nearer the phone). So regular cell phone use might cause the tumors in 0.001 percent of those who used them, or 1 extra person in every 100,000 in that group.

Would Maria Feychting stop her own children from using cell phones? Not at all: she would rather know where they were and be able to call them. She warned that the results were provisional, the study small, and quite different results might emerge once they looked at a larger sample. In fact, it was usually the case, she said, that apparent risks like this seemed to diminish with more evidence and bigger surveys. Two years later the worldwide research group looking into the health effects of cell phones—Interphone—of which the Karolinska Institute team was a part, did indeed produce another report drawing on new results from a much larger sample. It now said there was no evidence of increased risk of acoustic neuroma from cell phones, the evidence in the earlier study having been a statistical fluke, the product of chance, which in the larger study disappeared.

A great many percentage rises or falls, in health statistics, crime, accident rates, and elsewhere, are susceptible to the same problem and the same solution. Reports of risks could stick to counting people, as people instinctively do, keep percentages to a minimum and use natural frequencies. Press officers could be encouraged to do the same, and then we could all ask: how many *extra* people per 100, or per 1,000 might this risk affect?

Risk is one side of uncertainty. There is another, potentially as confusing and every bit as personal.

Imagine that you are a hardworking citizen on the edge, wide-eyed, sleep-deprived, while nearby in the moonlight someone you can barely now speak of in civil terms has a faulty car alarm. It's all right; we understand how you feel, even if we must stop short, naturally, of condoning your actions as you decide that if he doesn't sort out the problem, right now, you might just remember where you put the baseball bat.

The alarm tells you with shrill self-belief that the car is being broken into; you know from weary experience that the alarm in question can't tell theft from moped turbulence. You hear it as a final cue to righteous revenge; a statistician, on the other hand, hears a false positive.

False positives are results that tell you something important is afoot, but are wrong. The test was done, the result came in, it said "yes," but mistakenly, for the truth was "no." All tests have a risk of producing false positives.

There is also a risk of error in the other direction, the false negative. False negatives are when the result says "no," but the truth is "yes." You wake up to find, after an uninterrupted night's sleep, that the car has finally been stolen. The alarm—serves him right—had nothing to say about it.

There are a hundred and one varieties of false positive and negative; the cancer clusters in Chapter 3 on chance are probably a false positive, and it is in health generally that they tend to be a problem, when people have tests and the results come back saying they either have or haven't got one condition or another. Some of those test results will be wrong. The accuracy of the test is usually expressed as a percentage: "The test is 90 percent reliable." It has been found that doctors, no less than patients, are often hopelessly confused when it comes to interpreting what this means in human terms.

Gerd Gigerenzer is a psychologist, director of the Center for Adaptive Behavior and Cognition at the Max Planck Institute in Berlin. He asked a group of physicians to tell him the chance of a patient truly having a condition (breast cancer) when a test (a mammogram) that was 90 percent accurate at spotting those who had it, and 93 percent accurate at spotting those who did not, came back positive.

He added one other important piece of information: that the condition affected about 0.8 percent of the population for the group of forty- to fifty-year-old women being tested. Of the twenty-four physicians to whom he gave this information, just two worked out correctly the chance of the patient really having the condition. Two others were nearly right, but for the wrong reasons. Most were not only wrong, but hopelessly wrong. Percentages confused the experts like everyone else.

Quite a few assumed that, since the test was 90 percent accurate, a positive result meant a 90 percent chance of having the condition, but there was a wide variety of opinion. Gigerenzer comments: "If you were a patient, you would be justifiably alarmed by this diversity." In fact, more than nine out of ten positive tests under these assumptions are false positives, and the patient is in the clear.

To see why, look at the question again, this time expressed in terms that make more human sense, natural frequencies.

Imagine 1,000 women. Typically, eight have cancer, for whom the test, a fairly but not perfectly accurate test, comes back positive in seven cases. The remaining 992 do not have cancer, but remember that the test can be inaccurate for them, too. Nearly seventy of them will also have a positive result. These are the false positives, people with positive results that are wrong.

Now we can see easily that there will be about seventy-seven positive results in total (the true positives and the false positives combined) but that only about seven of them will be accurate. This means that for any one woman with a positive test, the chance that it is accurate is low and not, as most physicians thought, high.

The consequences, as Gigerenzer points out, are far from trivial: emotional distress, financial cost, further investigation, biopsy, even, for an unlucky few, unnecessary mastectomy. Gigerenzer argues that at least some of this is due to a false confidence in the degree of certainty conferred by a test that is at least 90 percent accurate. If positive tests were reported to patients with a better sense of their fallibility, perhaps by speaking of people, not percentages, it would ease at least some of the emotional distress. But how does that false confidence arise? In part, because the numbers used are not in tune with human instinct to count.

Uncertainty is a fact of life. Numbers, often being precise, are sometimes used as if they overcame it. A vital principle to establish is that many numbers will be uncertain, and we should not hold that against them. Even 90 percent accuracy might imply more uncertainty than you would expect. The human lesson here is that since life is not certain, and since we know this from experience, we should not expect numbers to be any different. They can clarify uncertainty, if used carefully, but they cannot beat it.

Having tried to curb the habit of overinterpretation, we need to restrain its opposite, the temptation to throw out all such numbers. Being fallible does not make numbers useless, and the fact that most of the positives are false positives does not mean the test is no good. It has at least narrowed the odds, even if with nothing like 90 percent certainty. Those who are positive are still unlikely to have breast cancer, but they are a little more likely than before they were tested. Those who are negative are now even less likely to have it than before they were tested.

So it is not that uncertainty means absolute ignorance, nor that the numbers offer certainty, rather that they can narrow the scope of our ignorance. This partial foretelling of fate is an extraordinary achievement. But we need to keep it in proportion, and we certainly need to get the right way around the degree of risk, whether likely or not. The overwhelming evidence is that we are more likely to judge this correctly if we use natural frequencies when we can and count people, as people do, rather than use percentages.

What generally matters is not whether a number is right or wrong—they are often wrong—but whether numbers are so wrong as to be misleading. It is standard practice among statisticians to say how wrong they think their numbers might be, though we might not even know in which direction—whether too high or too low. Putting an estimate on the potential size of the error, which is customarily done by saying how big the range of estimates needs to be before we can be 95 percent sure it covers the right answer (known as a confidence interval) is the best we can do by way of practical precaution against the number being bad. Though even with a confidence interval of 95 percent there is still a 5 percent chance of being wrong. This is a kind of modesty the media often ignores. The news often doesn't have time, or think it important, to tell you that there was a wide range of plausible estimates and that this was just one, from somewhere near the middle. So we don't know, half the time, whether the number is the kind to miss a barn door, enjoying no one's confidence, or if it is a number strongly believed to hit the mark.

We accuse statisticians of being overly reductive and turning the world into numbers, but statisticians know well enough how approximate and fallible their numbers are. It is the rest of us who perform the worst reductionism whenever we pretend the numbers give us excessive certainty. Any journalist who acts as if the range of uncertainty does not matter, and reports only one number in place of a spread of doubt, conspires in a foolish delusion for which no self-respecting statistician would ever fall. Statistics is an exercise in coping with, and trying to make sense of, uncertainty, not in producing certainty. It is usually frank in admission of its doubt, and we should be more willing to do the same.

If ever you find yourself asking, as you contemplate a number, "How can they be so precise?" the answer is that they probably can't, and probably weren't, but the reporting swept doubt under the carpet in the interests of brevity. If, somewhere along the line, the uncertainty has dropped out of a report, it probably will pay to find out what it had to say.

If we accept that numbers are not fortune tellers and will never tell you everything, but can tell you something, then they retain an astonishing power to put a probability on your fate. The presentation of the numbers might have left a lot to be desired, but the very fact that we can know—approximately—what effect drinking regularly will have on your chances of breast cancer, is remarkable. Picking out the effect of alcohol from all the other lifetime influences on your health is a prodigious undertaking and the medical surveys that make it possible are massive data-crunching exercises. Having gone to all that effort, it is a scandal not to put it to proper use and hear clearly what the numbers have, even if with modesty, to say.

SAMPLING

8

DRINKING FROM A FIRE HOSE

ounting is often bluff. It is, in truth, counting-lite. Hundreds of numbers printed and broadcast every day have routinely, but necessarily, skimped on the job.

To know if they're reliable, you need to know how they are gathered. But few know the shortcuts taken to collect even respected numbers: the size of the economy or trade, the profit companies make, how much travel and tourism there is, productivity, the rate of inflation, the level of employment . . . as well as controversial "counts" like Iraq war dead, HIV/AIDS cases, immigrants, and more.

None of these is a proper tally, though their ups and downs are the bread and butter of news. Instead, only a few of each are counted, assuming they are representative of the rest, then multiplied to the right size for the whole country.

This is the sample, the essence of a million statistics, like the poet's drop of water containing an image of the world in miniature—we hope. It is wonderful, when it works. But if the few that are counted don't mirror the others, the whole endeavor fails. So which few? Choose badly and the sample is skewed, the mirror flawed, and for a great many of the basic facts about us, our country and our economy, all that is multiplied is the size of the error.

HIV/AIDS cases constitute a global emergency, but are fundamentally impossible to count fully. In 2007, UNAIDS (the international agency with responsibility for the figures—and for coordination of the effort to tackle them) said the disease was still spreading, but revised its figures down. That's right, they went up to a lower number—from about 40 million to 33 million worldwide. Researchers conceded that their sampling (much of it in urban maternity clinics) had been biased. Pregnant women turn out to be a poor reflection of the prevalence of a sexually transmitted disease in the rest of the population because—who'd have guessed?—pregnant women have all had unprotected sex.

At least this sample was easy to collect. When the job is hard and a full count impossible, convenience matters, but it's a strangely warped mirror to the world that implies everyone is having sex, and unprotected sex at that. The new, improved methodology incorporates data from population-wide surveys, where available.

So although UNAIDS thinks the problem is still getting worse, the figures are better, and not for the first time. There have been several methodological hiccups, the total having hit 40 million in 2001, and risen but fallen ever since (that 2001 estimate has also been retrospectively revised, from 40 million to about 27 million), though there are wide margins of uncertainty around all these figures. And they are still contested—in both directions—some believing them too low, others still too high, and argument rages on.

The Chicago University economist Emily Oster, for example, has looked at mortality rates, rather than infection rates, to argue that far fewer people are dying than the UN figures would lead us to expect. (Incidentally, not all increases in the numbers are bad news. Sometimes these are people who in the past would have died but are now surviving longer, and that is why there are more of them.)

Nothing here undermines the conclusion that HIV/AIDS is a humanitarian disaster: it's believed to kill about 2 million people each year, with 2.5 million newly infected. In some places, earlier progress seems to be in reverse. Though certainly still wrong one way or another, these figures would have to be very wrong indeed to be anything less than calamitous.

But the example shows how apparently reliable numbers can stand on fallible methods, and how this risk is also unavoidable. So much to count, so little time; expensive, inconvenient, overwhelming, the effort to quantify even the important facts would be futile, but for sampling. The least we can do is understand how ubiquitous it is, and how it goes wrong.

Take immigration, controversial in the United States as in Britain. An informed debate would start with good data (though we'll see later that people often form opinions in extreme ignorance). And the United States does indeed devote huge effort to its Census data, mightily impressive to we outsiders, and once every ten years tries to count everyone. In between, to keep up with a rapidly changing demographic, it samples. One small example shows the potential pitfalls.

Each year in the big cities, the sample takes in different neighborhoods and buildings. At one stage of this process, several housing units in the same block or building are selected for interview. Since new immigrants tend to go where members of their ethnic group already live, these groups form clusters. In New York City for instance, many Jamaicans, Dominicans, and other immigrants from the Caribbean live in close proximity. Some years the sample chances on one of their clusters, some years it doesn't. Extrapolate from it and the Caribbean population appears to go up and down like a yo-yo.

Immigration statistics in Britain became, in the summer of 2007, a subject of ridicule, repeatedly revised, always up. Press, politicians, and presumably some members of the public seemed scandalized that they were a) inaccurate and b) not actually counted.

To see where the problems begin, follow us for a brief excursion across the English Channel. The numbers for sea crossings begin in the cold early morning with a huddle of officials at Dover docks on the English south coast. Their job is to count, and there's a lot at stake. The numbers are the crux of a passionate argument about whether the English way of life is imperiled and the country too crowded, or whether Britain needs more workers if it is to prosper. These officials count, or rather sample, migrants.

All the public normally hears and knows of the statistical bureaucracy at its borders is the single number that hits the headlines: in 2005 net immigration to Britain of about 180,000, or 500 a day. This is sometimes enriched with a little detail about where they came from, how old they are, whether single, with children, and so on.

Officials from the International Passenger Survey, on the other hand, experience a daily clash with compromising reality in which the supposedly simple process of counting people to produce those numbers is seen for what it is: mushiness personified. Here they know well the softness of people, people on the move even more so. What's more, they sample a tiny fraction of the total.

In matching blue blazers on a dismal gray day, survey teams flit to and from France, weaving between passengers on the ferries, from the duty-free counter to the truckers' showers, trying in all modesty to map the immense complexities of how and why people cross borders. The problem is that to know whether passengers are coming or going for good, on vacation or a booze cruise, or a year off from college starting in Calais and aiming for Rio, there's little alternative to asking them directly. And so the tides of people swilling about the world, seeking new lives and fleeing old, heading for work, marriage, or retirement in the sun, whether "swamping" Britain or "brain-draining" from it, however you interpret the figures, are captured for the record if they travel by sea when skulking by slot machines, halfway though a croissant, or off to the ladies' room.

"Hey, you in the Levi's! Yes, by the lifeboat! Where are you going?" And so they discover the shifting sands of human migration and ambition.

Or maybe not, not least because it is a lot harder than that. To begin with, the members of the International Passenger Survey teams are powerless—no one has to answer their questions—and, of course, they are impeccably polite; this is no job for the rude or impatient. Next, they cannot count and question everyone—there is not enough time—so they take a sample which, to avoid the risk of picking twenty people drinking their way through the same stag weekend and concluding the whole shipload is doing the same, has to be as random as possible.

So, shortly before departure, they stand at the top of the eight flights of stairs and various lifts onto the passenger deck as the passengers come aboard, scribbling a description of every tenth in the file; of the backpackers, the refugees, the suited, or the carefree, hoping to pick them out later for a gentle interrogation: tall guy, beard, "Surfers Do It Standing Up" T-shirt. That one is easy enough, not much likelihood of a change of clothes either, which sometimes throws a monkey wrench in the works.

When several hundred Boy Scouts came aboard en route to the World Scout Jamboree, the survey team assumed, with relief, that the way to tell them apart would be by the color of their neckerchief slides (known to boy scouts as "woggles"), until it turned out that the Jamboree had issued everyone with new, identical ones. The fact that they were all going to the same place and all coming back again did not absolve the survey teams of the obligation to ask. The risks of neckerchief fatigue are an occupational hazard of all kinds of counting.

The ferry heaves into its journey and, equipped with their passenger vignettes, the survey team members also set off, like naturalists in the undergrowth, to track down their prey, and hope they all speak English.

"I'm looking for a large lady with a red paisley skirt and blue scarf. I think that's her in the bar ordering a gin and lime."

She is spotted—and with dignified haste the quarry is cornered: "Excuse me, I'm conducting the International Passenger Survey. Would you be willing to answer a few questions?"

"Of course." Or perhaps, with less courtesy: "Nah! Too busy, luv." And with that the emigration of an eminent City financier is missed, possibly. About 7 percent refuse to answer. Some report their honest intention to come or go for good, then change their minds, and flee the weather, or the food, after three months.

This is counting in the real world. It is not a science, it is not precise—in some ways it is almost, with respect to those who do it, absurd. Get down to the grit of the way data is gathered, and you often find something slightly disturbing: human mess and muddle, luck and judgment, and always a margin of error in gathering only a small slice of the true total, from which hopeful sample we simply extrapolate. No matter how conscientious the counters, the counted have a habit of being downright inconvenient, in almost every important area of life. The assumption that the things we have counted are correctly counted is rarely true, cannot, in fact, often be true, and as a result our grip on the world through numbers is far feebler than we like to think.

The International Passenger Survey teams in Britain interview about 300,000 people a year, on boats and at airports. In 2005 about six or seven hundred of these were migrants, a tiny proportion from which to estimate a total flow in both directions of hundreds of thousands of people (though recently they began supplementing the routine survey data with additional surveys at Heathrow and Gatwick designed particularly to identify migrants). The system has been described by Mervyn King, governor of the Bank of England—who is in charge of setting interest rates and has good reason for wanting to know the size of the workforce—as hopelessly inadequate. In November 2006, he told a Parliamentary committee: "We do not have any particularly accurate method of measuring migration, either gross or net." He added: "In 2003, I think there were 516,000 passenger journeys between the UK and Poland. That is both in and out. Five hundred and five thousand of those—that is almost all of them—were to Gatwick, Heathrow, or Manchester (airports). Over the next two years the number of passenger journeys between the UK and Poland went from 516,000 to about 1.8 million. Almost all of that increase was in airports other than Heathrow, Gatwick, and Manchester. Why does this matter? Because most of the people handing out the questionnaires for the International Passenger Survey were at Heathrow, Gatwick, and Manchester. The number of people who declared themselves migrants and were sampled at airports outside the main three was just 79."

You have to laugh at this irresistible image: the International Passenger Survey enumerators looking one way while a million people tiptoe to and fro behind them. Whether they should have foreseen this change in the pattern of travel is still arguable, but it illustrates perfectly how new trends can leave old samples looking unfortunate.

It does not make this sample, or others, useless. Flawed as the data is, the attempt is usually better than no data at all; and improvement is seldom easy without expense. Let's be realistic—samples are often the best we can do in an infinite and imperfect world. But we should arm ourselves against the scale of their imperfection: typically less robust than expected, sometimes disconcertingly so, on occasion even comical. The key point is to realize the implicit uncertainty of these numbers and treat them with the care—not cynicism—they deserve.

Currently more controversial even than the figures for HIV/AIDS or immigration is the Iraq war. In October 2006 it was estimated in research by a team from Johns Hopkins University, published in the *Lancet*, that nearly twice as many people had died as a result of the Anglo-American invasion/liberation of Iraq as Britain had lost in World War II, about 650,000 in Iraq to about 350,000 British war dead, civilian and combatants combined; that is, nearly twice as many dead in a country with about half the population.

Of those 650,000 deaths, about 600,000 were thought to be directly due to violence, a figure that was the central estimate in a range from about 400,000 to about 800,000. The comparison with World War II shows that these are all extremely big numbers. The political impact of the estimate matched the numbers for size, and its accuracy was fiercely contested.

It was, of course, based on a sample. Two survey teams visited 50 randomly selected sites of about 40 households each, a total of 1,849 households with an average of about 7 members (nearly 13,000 people). One person in each household was asked about deaths in the fourteen months prior to the invasion and in the period after. In about 90 percent of cases of reported deaths they asked to see death certificates, and were usually obliged.

The numbers were much higher than those of the Iraq Body Count (which had recorded about 60,000 at that time), an organization that uses two separate media reports of a war death before adding to its tally (this is a genuine count, not a sample), and is scrupulous in trying to put names to numbers. But because it is a passive count, it is highly likely to be (and is, by the admission of those who do it) an undercount.

But was it likely to have been such a severe undercount that it captured only about 10 percent of the true figure? Attention turned to the sampling methodology of the bigger number. Because it was a sample, each death recorded had to be multiplied by roughly 2,200 to derive a figure for the whole of Iraq. So if the sample was biased in any way—if, to take the extreme example, there had been a bloody feud in one tiny, isolated, and violent neighborhood in Iraq, causing 300 deaths, and if all these deaths had been in one of the areas surveyed, but there had been not a single death anywhere else in Iraq, the survey still would produce a final figure of 650,000, when the true figure was 300.

Of course, the sample was nothing like as lopsided in this way, to this degree, but was it lopsided in some other? Did it, as critics suggested, err in sampling too many houses close to the main streets where the bombs and killings were more common, and not enough in quieter, rural areas? Was there any manner in which the survey managed the equivalent of counting Pearl Harbor but missing Alaska in a sample of America's war dead?

In our view, *if* the Iraq survey produced a misleading number (it is bound to be "wrong" in the sense of not being precisely right, the greater fault is to be "misleading"), it is more likely because of the kind of problem discussed in the next chapter—to do with data quality—than a bad sample. What statisticians call the "design" of the sample was in no obvious way stupid or badly flawed. But it is perfectly proper to probe that design, or indeed any sample, for weakness or bias.

A good case where we already know for sure that bias applies, but not what to do about it, is in a number capable from time to time of dominating the headlines: the growth of the economy.

The Federal Reserve, the Treasury, politicians, the whole business community, and the leagues of economic analysts and commentators generally accept the authority of the figures for U.S. economic growth, compiled in good faith with rigorous and scrupulous determination. It is a figure that has the administration trembling or rejoicing, the bedrock of every economic forecast, the measure of success of an economy, the record of rising prosperity or recession.

It is also based on a sample. What's more, there is good evidence the sample is systematically biased.

The problem is that the fastest-growing parts of the economy are often new, comprising businesses that simply weren't there before. If we counted every cent of economic activity as it took place, we'd know all about them, but we don't. Not until they file their returns to the Internal Revenue Service do we get a proper sense of where to find the latest boom. So in order to measure economic growth, what do we put in our sample of businesses? Businesses that we already know about, what else?

Oddly enough, this does not lead to an underestimate of growth, since the problem is known and understood. The figures are adjusted, with an estimate of growth in new areas. However, that estimate is usually too high, and initial reports of GDP growth in the United States are almost always revised down.

In the UK, by contrast, this new growth is simply ignored until the tax returns come in to tell us what it was. So the initial UK GDP figures fail to count the one area of the economy that is plausibly growing fastest—the new firms with the new ideas, creating new markets. This has led to initial underreporting of economic growth in the UK typically by about half a percentage point. When growth moves along at about 2.5 percent a year, that is a big error.

In consequence, for the last ten years the UK has believed itself underperforming compared with the United States, when in fact it might have been doing every bit as well.

By the time more accurate figures come out, of course, public attention has moved on, thinking the United States a hare and the UK a sloth. Not on this data, it's not. Our opinions have been shaped by our sampling, and our sampling is consistently biased in a way that is hard to remedy accurately until later. When the figures first came out, we were told that growth in 2004—the latest year for which we have three-year revisions—was 2.2 percent in the UK (not bad, but not great) compared to U.S. growth of 3.2 percent (impressive). Now it is believed UK growth was 2.7 percent, and U.S. growth also 2.7 percent.

What's needed to spot a spanner in the sampling works, above all, is imagination—that and a thirst for detail. This isn't an especially technical exercise; it's a feel for practicalities. What kind of bias might be lurking? What are the peculiarities of the sample that mean the few we happen to count will be unrepresentative of the rest? For a test of your own imagination, try hedgehogs.

The National Hedgehog Survey began in Britain in 2001. Less cooperative even than the Census-shy public, hedgehogs keep their heads down. In the wild there is no reliable way to count them short of rounding them up, dead or alive, or ripping through habitats, ringing legs—which would somewhat miss the point.

It is said hedgehogs are in decline. How do we know? We could ask gamekeepers how many they have seen lately—up or down from last year—and we could put the same questions in general opinion surveys. But the answers would be impressionistic, anecdotal, and we want more objective data than that. What do we do?

In 2002 the Hedgehog Survey was broadened to become the Mammals on Roads Survey. Its name suggests, as young hedgehog lovers might say, an icky solution. The survey is in June, July, and August each year, when mammals are on the move and when, to take the measure of their living population, we count how many are silhouetted on the tarmac. The more hedgehogs there are in the wild, so the logic goes, the more will snuffle to their doom on the bypass. In a small hog population this will be rare, in a large one more common.

So here's the test: what kind of bias might creep into such a sample. You'll find some suggestions at the end of the chapter.

Life as a fire hose, and samplers with china teacups and aristocratically crooked fingers, is an unequal statistical fight. In truth, it is amazing we capture as much as we do, as accurately as we do. But we do best of all if we also recognize the limitations, and learn to spot occasions when our sample is most likely to miss something.

Like when it's small, for example. How many hours—in total—do American parents read with their children in the five or more years before starting school? If they are poor parents, the average is apparently just twenty-five. If they are "middle income" parents, the number is between 1,000 and 1,700 hours. You might have seen these figures before. You might be tempted to think they explain a thing or two.

They first appeared in 1990 in a book called *Beginning to Read: Thinking and Learning About Print* by Marilyn Jager Adams. Naturally, she did not count the reading hours in every family—how could she? She sampled them, and devoted five pages to an honest account of how she did it.

The twenty-five-hour estimate derived from figures reported in a separate study in 1986 of twenty-four children in twenty-two low-income families in Southern California—a tiny sample—and its author reportedly has advised against using it to make generalized claims.

The middle-income sample, however, is even smaller. It is one. The one in question was Marilyn Jager Adams's son, John.

And yet, according to Carl Bialik, author of the "Numbers Guy" column in the *Wall Street Journal* (and to whom we owe this example), in the seventeen years since her book, "at least a half-dozen child-advocacy groups, including United Way, Kids in Common, and Everybody Wins, have boiled down those five pages into a single sentence, repeated in various forms, often without attribution to the original source. As is typical for such numbers, the child-reading stats have taken on a life of their own through a game of media telephone, with news articles usually attributing the numbers to one of these advocacy groups or to various researchers or foundations that themselves got the numbers from the Adams book."

Is John typically middle income? As Carl Bialik says: "That's akin to predicting that all young children from middle-income families will graduate college with a degree in psychology and statistics, as John has done."

Even data that seems to describe us personally is prone to bias. You know your own baby, surely. But do you know if junior is growing properly? The simple but annoying answer is that it depends on what "properly" means. This is defined in a booklet showing charts that plot the baby's height, weight, and head circumference, plus a health visitor sometimes overly insistent about the proper place on the chart for your child to be. The fact that there is variation in all children—some tall, some short, some heavy, some light—is small consolation to the fretful parent whose child is below the central line on the graph, the fiftieth percentile. In itself, such a worry is usually groundless, even if encouraged. All the fiftieth percentile means is that half of babies will grow faster and half slower. It is not a target, or worse, a test that children pass or fail.

But there is another problem: who says this is how fast babies grow? On what evidence? The evidence, naturally, of a sample. Who is in that sample? A variety of babies past that were supposed to represent the full range of human experience. Is there anything wrong with that?

Yes, as it happens. According to the World Health Organization, not every kind of baby should have a place in the sample. The WHO wants babies breast-fed and says the bottle-fed should be excluded. This matters because bottle-fed babies tend to grow more quickly than breast-fed babies, to the extent that after two years the breast-fed baby born on the fiftieth percentile line will have fallen to the bottom 25 percent of the current charts.

If the charts were revised, as the WHO wants, formerly average-sized, bottle-fed babies would start to look big, while smaller, breast-fed babies would start to look average.

So the WHO thinks the charts should set a norm—asserting the value of one feeding routine over another, and has picked a sample accordingly. This is a justifiable bias, it says, against bad practice, but a bias nevertheless, taking the chart away from description and toward prescription, simply by changing the sample.

All kinds of characteristics can, and do, cause bias. Did the sample somehow pick up more people who were older, younger, married, unemployed, taller, richer, fatter; were they more or less likely to be smokers, car drivers, female, parents, left-wing, religious, sporty, paranoid . . . than the population as a whole, or any other of the million and one traits that distinguish us, and might just make a difference?

In one famous case it was found that a sample of Democrat voters in the United States were less satisfied with their sex lives than Republicans, until someone remembered that women generally reported less satisfaction with their sex lives than men, and that more women tended to vote Democrat than Republican.

Bias can seep into a population sample in as many ways as people are different, in their beliefs, habits, lifestyles, history, biology. Bias of some kind, intentional, careless, or accidental, is par for the course in many samples that go into magazine surveys, or surveys that serve as a marketing gimmick, and generally get the answer they want, if only because they are completed by people who are already interested in the subject of the survey. Bias is less likely in carefully designed surveys that aim for a random sample of the population. But surveys asking dumb questions and getting dumb answers are not alone in finding that potential bias lurks everywhere, with menace, trying to get a foot in the door, threatening to wreck the integrity of our conclusions.

A good hint that a count is less than it seems is the kind of big, round number supposedly summing the experience of the nation, as happened with a virus in Britain in winter 2007. There were stark warnings—don't touch the door handles, don't shake hands, don't go out, scrub and scrub again—and lurid images: of sickness, deserted workplaces, closed hospital wards. The numbers were huge.

The British media was in the grip of an epidemic, its pages covered in vomit, at least reports of it. The condition was first sighted in the *Daily Telegraph*, so virulent that within days every other newspaper and broadcaster had succumbed, sick with the same vile imaginings.

In these dire circumstances, only one cure is possible: sufferers are strongly advised to check the sources of their data. Strangely, none did, so the small matter of whether the public itself faced a real epidemic became almost irrelevant.

It was the winter of 2007 to 2008 and the sickness was norovirus, also known as winter flu or winter vomiting disease, and it seemed that an alarming number of people had fallen victim. The *Daily Telegraph* newspaper said 3 million. The *Daily Express* said 3.5 million. The *Sun* said 4 million. From those bold numbers, you might expect more officials with clipboards, this time stationed outside every bathroom window in the land, recording how many were throwing up.

But clearly the number of cases is not counted in any proper sense of the word. Only a tiny proportion of those sick with norovirus go to the doctor. Fewer still are confirmed with a lab test. Norovirus passes (there is no cure) in a couple of days. In truth, no one could know (nor did) how many were affected, but they arrived at alarming totals on the basis of a miniscule sample from which they simply extrapolated.

Samples have to be large enough to be plausibly representative of the rest of the population. The sample on this occasion—the only data we have about the incidence of norovirus, in fact—was the 2,000 cases occurring in October, November, and December of 2007 that had been confirmed in the laboratory. From 2,000 confirmed cases to 3 or 4 million is a big leap, but, for every recorded case, the media reported that there were 1,500 more in the community, a figure obtained from the Health Protection Agency (HPA). This made the arithmetic apparently straightforward: 2,000 confirmed cases, $\times$ 1,500 = 3 million.

But the HPA also urged caution, saying the ratio of confirmed cases to the total should not be taken literally, and might be changing anyway as, for example, diagnostic technology became more sensitive. People's readiness to go to the doctor with this illness also might have been changing. So the ratio of 1 to 1,500 is unreliable from the start; how unreliable, we'll now find out.

It originated in what is known as the IID (infectious intestinal diseases) Study conducted between 1993 and 1996. Such studies normally are careful about the claims they make. Researchers recognize that there is a good deal of uncertainty around the numbers they find in any one community, and that these may vary from place to place and time to time. And so they put around the numbers what they call confidence intervals. As the name suggests, these aim to give some sense whether this is the kind of number that has their confidence, or one they wouldn't trust to hit a barn door. A rough rule of thumb is that wide confidence intervals indicate that the true figure is more uncertain, narrow confidence intervals suggest more reliability.

The estimate, remember, was that every case recorded equaled "about" 1,500 in total. So how "about" is "about"? How wide were the confidence intervals for the norovirus? To be 95 percent sure of having got it right, the confidence intervals in the IID Study said that 1 lab case might be equal to as few as 140 cases in the community . . . or as many as 17,000. That's some uncertainty.

These numbers imply that for the 2,000 cases confirmed in the laboratory in the winter of 2007, the true number in the community could be anywhere between 280,000 and 34 million (more than half the entire population of the UK), with a 5 percent chance, according to this research, that the true value lies outside even these astonishingly wide estimates. As the authors of the study said when reporting their findings in the restrained language of the *British Medical Journal*: "There was considerable statistical uncertainty in this ratio." Let's put it more bluntly: they hadn't the foggiest idea, and for good reason—the number of laboratory cases in that study, the number of people whose norovirus had been confirmed, and thus the sample from which the 1 to 1,500 ratio was calculated, was in fact . . . 1.

This was a "wouldn't-hit-a-barn-door-number" if ever there was one. What's more, three quarters of the 2,000 recorded cases in the winter of 2007 "epidemic" were from patients already in hospital wards, where the illness is known to spread quickly in the contained surroundings of a relatively settled group of people. That is, these cases, the majority, might not have been representative of more than a handful of others in the outside world, if any.

By mid-January 2008, all talk of an epidemic had disappeared, as the apparent rise in cases evaporated (2,000 in the last three months of 2007 had been twice the corresponding period for the previous year, but there weren't really twice as many recorded cases; the peak just had arrived earlier—and also passed earlier—making the seasonal total much like any other). None of this makes much difference to the more fundamental problem—that we have only the vaguest idea how the sample relates to the rest of the population, and none at all about whether that relationship is changing, and are therefore unable to say—without some other source of overwhelming evidence—whether there's a comparative torrent of vomit, or a trickle.

It is tempting to give up on all things sampled as fatally biased and terminally useless. But bias is a risk, not a certainty. Being wise to that risk is not cynicism—it is what any statistician worth their salt strives for.

And sampling is, as we have said, inevitable: there is simply too much to count to count it properly. Sometimes the sample can be invaluable.

Take the extreme example of an uncountable number that we nevertheless want to know: how many fish are there in the sea?

If you believe the scientists, not enough; the seas are empty almost to the point where some fish stocks cannot replace themselves. If you believe the fishermen, there is still a viable industry. Whether that industry is allowed to continue depends on counting those fish. Since that is impossible, the only alternative is sampling.

This was true off the coast of Newfoundland, where for several years the scientific advice was to curtail the catch until, in 1992, the stocks of cod catastrophically collapsed, leading to the overnight death of an industry and the loss of tens of thousands of jobs.

It probably is true in Georges Bank, off the coast of Martha's Vineyard, Massachusetts, a vast fishing ground where the industry has resisted sampling evidence that the stocks are under severe pressure, insisting the quantity of fish is recovering.

When *More or Less* visited the daily fish market in Newlyn, Cornwall, in 2005, to look at the evidence of depleted stocks in the North Sea, the consensus there was that the catch was bigger than twelve to fifteen years ago. "We don't believe the scientific evidence," they said.

The scientists had sampled cod fish stocks at random and weren't finding many. The fishing industry thought them stupid. One industry spokesperson told us, "If you want to count sheep, you don't look at the land at random, you go to the field where the sheep are." In other words, the samplers were looking in the wrong places. If they wanted to know how many fish there were, they should go where there were fish to count. When the fishermen went to those places, they found fish in abundance.

The International Council for the Exploration of the Sea recommends a quota for cod fishing in the North Sea of zero. Fisheries ministers from the European Union countries regularly ignore this advice during what has become an annual feud about the right levels of TACs (total allowable catches) and in 2006 the EU allowed a catch of about 26,000 tons.

Is the scientists' sampling right? In all probability, it is—by which we mean, of course, that it is not precisely accurate, but nor, more importantly, is it misleading. Whereas fishing boats might be sailing for longer or farther to achieve the same level of catch, research surveys will not. They trawl for a standard period, usually half an hour. Nor will they upgrade their equipment, as commercial fishers do, to maximize their catch. And on the question of whether they count in the wrong places, and the superficial logic of going where the fish are, this ignores the possibility that there are more and more places where the fish are not, and is itself a bias. Going to where the fish are in order to count them is a little like drawing the picture of the donkey *after* you have pinned up the tail.

The fishermen, like all hunters, are searching for their prey. The fact that they find it and catch it tells us much more about how good they are at hunting than about how many fish there are. The scientists' samples are likely to be a far better guide to how many fish there are in the sea than the catch landed by determined and hardworking fishermen.

The kind of sampling that spots all these red herrings has to be rigorous and imaginative for every kind of bias. After all that, the sampling still might not be perfect, but it has performed a near miracle.

Finally, back to the hedgehogs, and the question of whether the sample of roadkill tells us what's happening to the population as a whole. One possible glitch is whether it counts hedgehogs or traffic density. Even if the hedgehog population was stable, more cars would produce more squashes. Or, as the more ingenious listeners to *More or Less* suggested, does the decline in roadkill chronicle the evolution of a smarter, traffic-savvy hedgehog which, instead of rolling into a ball at sight or sound of hazard, now makes a getaway, and lives to snuffle another day beyond the knowledge of our survey teams? Or maybe climate change has led to alterations in the hedgehog life cycle through the year, which reduces the chance of being on the roads in the three monitored months.

The most recent Mammals on Roads Survey tells us that average counts in both England and Wales have fallen each year of the survey, and the numbers in Scotland last year are markedly lower than those at the start of the survey. In England, the decline is biggest in the Eastern and East Midlands region and in the Southwest; no one knows why.

The lesson of the survey is that we often go to bizarre but still unsatisfactory lengths to gather enough information for a reasonable answer to our questions, but that it will remain vulnerable to misinterpretation, as well as to being a poor sample. Yet, for all the potential flaws, this really is the best we can do in the circumstances. Most of the time we don't stop to think, "How did they produce that number?" We have been spoiled by the ready availability of data into thinking it is easily gathered. It is anything but. We should not assume there is an obvious method bringing us a sure answer. We rarely know the whole answer, so we look for a way of knowing part, and then trust to deduction and guesswork. We take a small sample of a much larger whole, gushing with potential data, and hope: we try to drink from a fire hose.

In fact, drinking from a fire hose is relatively easy compared with statistical sampling. The problem is how you can be sure that the little you can swallow is like the rest. If it's all water, fine, but what if every drop is in some way different?

Statistical sampling struggles heroically to swallow an accurate representation of the whole, and often fails. What if the sample of flattened hedgehogs comes entirely from a doomed minority, out-evolved and now massively outnumbered by runaway hedgehogs? If so, we would have a biased count—of rollers (in decline) instead of runners (thriving). Unlikely, no doubt, and completely hypothetical, but how would we know?

DATA

9

KNOW THE UNKNOWNS

Asked about the facts and figures of American life, college students are often so wrong—and find the correct answers so surprising—that some change their minds about policy on the spot.

In Britain, the most senior civil servants, who implement and advise on policy, are often clueless about basic numbers on the economy or society.

Strong views and serious responsibilities are no guarantee of even passing acquaintance with the data. But numbers do not fall ripe into our laps. Someone has to find and fetch them. Far easier, some feel, not to bother, and trust their prejudices instead. Too bad that ignorance is an unreliable foundation for forthright opinion, let alone policy. For though much of what can be known through numbers is foggy, a fair portion of what people do not know is due to complacency, sloppiness, or fear. Policies have been badly conceived, even harmful, for want of looking up the obvious numbers in the department next door. People have died needlessly in the hospital for want of a tally to tell us too many had died already. Many of us build passionate beliefs on numbers that are figments of our imagination.

The deepest pitfall with numbers owes nothing to numbers themselves and much to the slack way they are treated, with carelessness all the way to contempt. But numbers, with all the caveats, are potent and persuasive, a versatile tool of understanding and argument. Treat them well and they will reward you. They are also often all we have got, and ignoring them is a dire alternative. For those at sea with the numbers, all this should be strangely reassuring. First, it means you have illustrious company. Second, it creates the opportunity to get ahead. Simply show enough care for numbers' integrity to think them worth treating seriously, and you are well on the way to empowerment.

Michael Ranney likes asking questions. Being an academic at UCLA, he has plenty of informed and opinionated young people for guinea pigs, but he keeps the questions straightforward. Even so, he doesn't expect the students to know the answers accurately; his initial interest is merely to see if they are in the right ballpark. For example: for every 1,000 U.S. residents, how many legal immigrants are there each year? How many people, per thousand residents, are incarcerated? For every thousand drivers, how many cars are there? For every thousand people, how many computers? How many abortions are there? How many murders per million inhabitants? And so on.

Few of us spend our leisure hours looking up and memorizing data. But many of us flatter ourselves that we know about these issues. And yet . . .

On abortion and immigration, says Ranney, about 80 percent of those questioned base their opinions on significantly inaccurate base-rate information. For example, students at an elite college typically estimated annual legal immigration at about 10 percent of the existing population of the United States (implying that for a population of 300 million, there were 30 million legal immigrants every year). Others—nonstudents—guessed higher.

The actual rate was about 0.3 percent. That is, even the lower estimates were more than thirty times too high. If your authors made the numerically equivalent mistake of telling you they were at least 165 feet tall, it's unlikely you'd take seriously their views on human height, or anything else. You probably would advise them, for their own good, to shut up.

The students' estimates for the number of abortions varied widely, but the middle of the range was about 5,000 for every million live births. The actual figure in the United States at that time (2006) was 335,000 per million live births—that is sixty-seven times higher than the typical estimate. These answers, in the famous words of the physicist Wolfgang Pauli, are not only not right, they are so far off the mark as to be not even wrong.

The next step is equally revealing. When the students found out the true figures, it made a difference. More thought there should be a big decrease in abortions. Among those who initially thought that abortions should always be allowed, there was movement toward favoring some restriction. We make no comment about the rightness or otherwise of these views. We simply observe the marked effect on them of data.

Professor Ranney, from the University of California's School of Education, says that if people are first invited to make their own guesses and are then corrected, their sense of surprise is greater than if they are simply given the correct figure at the outset. And surprise, it turns out, is useful. It helps to make the correct figure more memorable, and also makes more likely a change in belief about policy.

For our purposes, the lesson is simpler. Many educated people voicing opinions on the numbers at the heart of social and economic issues in truth haven't the faintest idea what the numbers are. But— and it is a critical but—they do change their thinking when they find out.

One more example from Michael Ranney's locker: unexpected numerical feedback on death rates for various diseases led under- graduates to provide funding allocations that more closely tracked those rates. Initially, they tended to overestimate the incidence, for example, of breast cancer compared with heart disease—and allo- cated a notional $100 accordingly. Once they knew the figures, they moved more money to heart disease. This fascinating work seems to make a strong case, contrary to the skeptical view, that opinions are not immune to data, but rather that accurate data does matter to people.

The alternative to using data—to rely instead on hunch or prejudice—seems to us indefensible, if commonplace. For example, at frequent talks and seminars over the past ten years or more, we have played a similar game in Britain, asking the country's most senior civil servants, journalists, numerous businesspeople, and academics a series of multiple-choice questions about very basic facts to do with the economy and society. Some, given their status or political importance, asked to remain anonymous. It is just as well they did.

Here is a sample of the questions, along with the answers given by a particular group of between 75 and 100 senior civil servants in September 2005. It would be unfair to identify them, but reasonable to say you certainly would hope that they understood the economy. (There is some rounding so the totals do not all sum to 100 percent, and not everyone answered all the questions.)

We have reframed the questions for the United States, using U.S. data immediately after each UK example so you can test your own knowledge.

What joint income (after tax) would a childless couple need to earn in order to be in the top 10 percent of earners?

UK OPTIONS . . .	% OF UK CIVIL SERVANTS WHO GAVE EACH ANSWER
A) £35,000	10
B) £50,000	48
C) £65,000	21
D) £80,000	19
E) £100,000	3
U.S. OPTIONS . . .	
A) $130,000	
B) $180,000	
C) $230,000	
D) $280,000	
E) $330,000	

The answer is £37,000 in the UK, and $130,000 in the United States in 2007. Some will resist believing that $130,000 or £37,000 after tax is enough to put a couple (both incomes combined—if they have two incomes) in the top 10 percent of the incomes in their respective countries. It is a powerful, instructive number and it is worth knowing. But the proportion of the UK group that did was just 10 percent. The single most common answer (£50,000) was nearly half as high again as it should have been. And 90 percent of a group of about seventy-five people, whose job it is to analyze our economy and help to set policy, think people earn far more than they really do, with more than 40 percent of them ludicrously wrong.

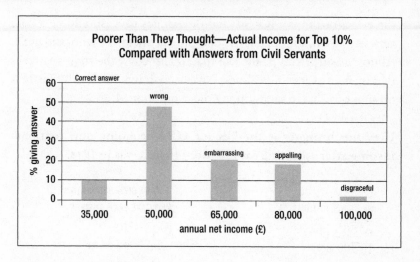

Poorer Than They Thought—Actual Income for Top 10% Compared with Answers from Civil Servants

What share of the income tax paid in the UK (or United States) is paid by the top 1 percent of earners?

UK OR U.S. OPTIONS . . .	% OF UK CIVIL SERVANTS WHO GAVE EACH ANSWER
A) 5% of all tax	19
B) 8%	19
C) 11%	24
D) 14%	19
E) 17%	19

Answer: They were all wrong. The correct answer is that the top earners pay 21 percent of all the income tax collected. (In the United States, the top 1 percent of earners pay about 35 percent of all income tax collected.)

It might seem unfair not to have given them the chance of getting it right, so it is reasonable to give credit to those who picked the biggest number available, 17 percent. All the others did poorly and almost two thirds thought the answer was 11 percent or less. Analyzing the effect of the tax system, and of changes to it, should be a core function of this group, but they simply did not know who paid what. Almost as surprising as the fact that so few knew the right answer is that their answers could almost have been drawn randomly. There is no sign of a shared view among those questioned.

How much bigger is the UK (or U.S.) economy now (national income after adjusting for inflation) than it was in 1948?

UK OR U.S. OPTIONS . . .	% OF UK CIVIL SERVANTS WHO GAVE EACH ANSWER
A) 50% bigger	10
B) 100%	25
C) 150%	42
D) 200%	17
E) 250%	5

The right answer is that the economy now is around 300 percent bigger than it was in 1948, and in the United States the equivalent number is around 600 percent (the United States population has grown much faster than that of the United Kindgom). This is another question where the right answer was not an option. But since only 5 percent of the group chose the highest possibility, it seems that most had no sense whatsoever of where the right answer lay. The UK economy grew in this period, on average, by about 2.5 percent a year. More than three quarters of the group gave answers that, if correct, would have implied half as much or less, and that we were less than half as well off as we actually are. That is quite an error— economics does not get much more fundamental than how fast the economy grows—and something of a shock.

There are 780,000 single parents on public assistance in the UK. How many are under age eighteen?

UK OPTIONS . . .	% OF UK CIVIL SERVANTS WHO GAVE EACH ANSWER
A) 15,000	29
B) 30,000	21
C) 45,000	0
D) 60,000	21
E) 75,000	29

The correct figure (for 2005) was 6,000. Those who chose the lowest option once again deserve credit. But there seems to be a common belief in this group and elsewhere that we have an epidemic of single teenage moms—a familiar political target—with half our group believing the problem at least ten times worse than it is, and some who no doubt would have gone higher had the options allowed.

The performance on these and other multiple-choice questions over the years from all sorts of groups has been unremittingly awful. That matters: if you want even the most rudimentary economic sense of what kind of a country you live in, it is hard to imagine it can be achieved without knowing, at least in vague terms, what typical incomes are. If you expect to comment on the burden of taxation, what chance a comment worth listening to if you are wholly in the dark about where the tax burden falls?

Numeracy requires an ability to use numbers properly. It also requires a proclivity for using them at all. The list of excuses for ignorance is long. It is far easier to mock than search for understanding, to say numbers do not matter, or they are all wrong anyway so who cares, or to say that we already know all that is important.

.   .   .

Wherever such prejudices take root, the consequences can be disastrous. When Joshua Loveday, aged eighteen months, was admitted for an operation at Bristol Royal Infirmary, in the west of England, no one told his parents that his chances of dying on the operating table at this hospital were significantly greater than elsewhere. Although some seemed to have had their suspicions, no one knew the measure of the extra risk Joshua faced.

He had been born with the main arteries to his heart the wrong way around. The operation to correct this—an arterial switch—is complicated, and it was to become clear that the surgeon had not mastered it.

"There were no doubts," said Joshua's mother, Mandy, or at least none she knew of. "If there had been any doubts then I would have been up and gone." But at 7:30 PM on January 12, Joshua died on the operating table.

His death began a series of investigations into what became a scandal. It was later established by an inquiry under Professor Sir Ian Kennedy that, of children operated on for certain heart conditions at Bristol, fully twice as many died as the national norm. It was described as one of the worst medical crises ever in Britain.

The facts began to come to light when an anesthetist, Dr. Steve Bolsin, arrived at Bristol from a hospital in London. Pediatric heart surgery took longer than he'd been used to; patients were on heart bypass machines a long time—with what effect he decided to find out, though he already suspected that death rates were unusually high. So he and a colleague rooted through the data where they discovered, they thought, persuasive evidence of what the medical profession calls excess mortality.

At first slow to respond, the hospital found itself—with Joshua's death the catalyst—eventually overwhelmed by suspicion and pressure for inquiries from press and public. The first of these was by an outside surgeon and cardiologist, another by the General Medical Council (the longest investigation in its history), and finally a third by the independent team led by Sir Ian Kennedy, which concluded that between thirty and thirty-five children probably had died unnecessarily.

Most of those involved believed they knew the effectiveness of their procedures and believed them to be as good as anyone's. None, however, knew the numbers; none knew how their mortality rates compared.

One crucial point attracted little attention at the time—grief and anger at the raw facts understandably swept all before them—but members of the inquiry team judged that had the excess mortality been not 100 per cent (twice as bad as others), but 50 percent, it would have been hard to say for sure that Bristol was genuinely out of line. That is, if about fifteen to seventeen babies had died unnecessarily rather than the estimated thirty to thirty-five, it might have been impossible to conclude that anything was wrong. Fifty percent worse than the norm strikes most as a shocking degree of failure, especially where failure means death. Why did mortality have to be 100 percent worse before the inquiry team was confident of its conclusions?

The two surgeons who took the brunt of public blame for what happened argued that, even on the figures available, it was not possible to show they had performed badly (and the inquiry itself was reluctant to blame individuals rather than the system at Bristol as a whole, saying that "The story of [the] pediatric cardiac surgical service [in] Bristol is not an account of bad people. Nor is it an account of people who did not care, nor of people who willfully harmed patients"). Mortality 100 percent worse than the norm was a big difference given the number of children involved, large enough to constitute "one of the worst medical crises" in Britain's history, but even then the conclusion was disputed.

The simple questions to establish the truth ought to be easy to answer: How many operations were there? How many deaths? How does that compare with others? Simple? The inquiry took three years.

Audrey Lawrence was one of the Bristol inquiry team, an expert in data quality. We asked her about the attention given to proper record keeping and the data quality for surgeon performance at Bristol. First, who kept the records?

"We found that we could get raw data for the UK cardiac surgeons register collected on forms since 1987, stored in a doctor's garage. It was nothing to do with the Department of Health, that's the point. The forms were collected centrally by one doctor, out of personal interest, and he entered the data in his own records and kept it in box files in the garage." There was no other central source of data about cardiac operations and their outcomes.

Next, how reliable was that data?

"My own experience of gathering data in hospitals was that it probably was not going to be accurate. We were very concerned about the quality of the data. All we had were these forms, and so we went round the individual units to see what processes had been followed. We found, as we suspected, that there was considerable lack of tightness, that there was a great deal of variety in the way data was collected, and a lot of the figures were quite suspect. It was quite a low priority [for hospitals]; figures were collected in a rush to return something at the end of the day."

So what, if any, conclusions could be drawn?

"The findings were fairly consistent that mortality in Bristol really did seem to be an outlier—something that doesn't fit the mold—with excess mortality of 100 percent, but had it been in the region of 50 percent, the quality of the data was such that we could not have been confident that Bristol was an outlier. We were sure, but only because the numbers were so different."

And did that conclusion mean there could be more places like Bristol with excess mortality, if only of 50 percent, that it would be difficult to detect?

"Undoubtedly."

It is disconcerting, to say the least, that attention to data quality can be so deficient that we still lack the simplest, most obvious, most desirable measure of health-care quality—whether we live or die when treated—to an acceptable degree of accuracy. Why, for so long, has it been impossible to answer questions like this? The answer, in part, is because the task is harder than expected. But it is also due to a lack of respect for data in the first place, for its intricacies and for the care needed to make sense of it. The data is often bad because the effort put into it is grudging, ill-thought-through, and derided as so much pen-pushing and bean-counting. It is bad, in essence, often because we make it so.

To see why data collection is so prone to misbehave, take an example of a trivial glitch in the system, brought to us by Professor David Hand of Imperial College, London. An e-mail survey of hospital doctors found that an infeasible number of them were born on November 11, 1911. What was going on?

It turned out that many could not be bothered to fill in all the boxes on the computer and had tried, where it said DoB, to hit 00 for the day, 00 for the month, and 00 for the year. Wise to that possibility, the system was set up to reject it, and force them to enter something else. So they did, and hit the next available number six times: 11/11/11; hence the sobering discovery that Britain's health service was chock-full of doctors over the age of ninety.

Try to measure something laughably elementary about people—their date of birth—and you find they are an awkward lot: tired, irritable, lazy, resentful of silly questions, convinced that "they"—the askers of those questions—probably know the answers already or don't really need to; inclined, in fact, to any number of other plausible and entirely normal acts of human awkwardness, any of which can throw a wrench in the works. Awareness of the frailty of numbers begins with a ready acknowledgment of the erratic ways of people.

Professor Hand said to us: "The idealized perception of where numbers come from is that someone measures something, the figure's accurate and goes straight in the database. That is about as far from the truth as it's possible to get."

So when forms from the 2001 Census in the UK were found in bundles in the trash can, or dumped on the doormat by enumerators at the end of a bad day of hard door-knocking and four-letter hospitality, or by residents who didn't see the point; when there was an attempted sabotage of questions on religious affiliation through an e-mail campaign to encourage the answer "Jedi Knights" (characters from the film *Star Wars*); when some saw the whole exercise as a big-brother conspiracy against the private citizen and kept as many details secret as they could, when all this and more was revealed as a litany of scandalous shortcomings in the numbers, what, really, did we expect, when numbers are so casually devalued?

The mechanics of counting are anything but mechanical. To understand numbers in life, start with flesh and blood. It is people who count, one of whom is worried her dog needs the vet, another dreaming of his next date, and it is other people they are often counting. What numbers trip over, often as not, is the sheer, cussed awkwardness and fallibility of us all. These hazards are best known not through obscure statistical methodology, but sensibility to human nature. We should begin by tackling our own complications and frailties, and ask ourselves these simple questions: "Who counted?" "How did they count?" "What am I like?"

One of the less noticed features of the numbers flying about in public life is how many critical ones are missing, and how few are well known. One of the most important lessons for those who live in terror of numbers, fearing they know nothing, is how much they often share with those who purport to know a lot.

In the case of patient records, the flow of data creates huge scope for human problems to creep in. Each time anyone goes to the hospital, there's a note of what happens. This note is translated into a code for every type of procedure. But the patient episode may not fit neatly into the available codes—people's illnesses can be messy after all: they arrive with one thing, have a complication, or arrive with many things and a choice has to be made of which goes on the form. Making sure that the forms are clear and thorough is not always a hospital priority. There are often gaps. Some clinicians help the coders decipher their notes, some don't. Some clinicians actually are hostile to the whole system. Some coders are well trained, some aren't. Although all hospitals are supposed to work to the same codes, variations creep in: essentially, they count differently. The coded data is then sent through about three layers of NHS bureaucracy before being published. It is not unusual for hospitals looking at their own data once it has been through the bureaucratic mill to say they don't recognize it.

Since Bristol, there are now more systems in place to detect wayward performance in the NHS. But are they good enough to rule out there still being centers with excess mortality that we fail to detect? No, says Audrey Lawrence, they are not.

And that is not the limit of the consequences of this difficulty with data collection in the health service. The NHS in England and Wales is introducing a system of choice for patients. In consultation with our GP, we are promised the right to decide where we want to be treated, initially from among five hospitals, and in time from any part of the health system.

Politicians have been confident that the right data can make it easier to choose where to go for the best treatment. Alan Milburn, then secretary of state for health, said, "I believe open publication will not just make sure we have a more open health service, but it will help to raise standards in all parts of the NHS." John Reid, also when secretary of state for health, said, "The working people of this country *will* have choice. They *will* have quality information. They *will* have power over their future and their health whether you or I like it or not."

The crunch comes with the phrase "quality information." Without quality information, meaningful choice is impossible. How do we know which is the best place to be treated? How do we know how long the wait will be? Only with comprehensive data comparing the success of one doctor or hospital with another, one waiting list with another. More often than not, this data is quantified.

At the time of writing, several years on from Mr. Milburn, and after Mr. Reid, too, has moved to another job, the best that Patient Choice can offer by way of quality information is to compare hospital car parking and cafeterias; on surgical performance, nothing useful is available to the public. There is one exception, though this is not routinely part of Patient Choice. In heart surgery, surgeons have set up their own Web site with a search facility covering every cardiac surgeon in the country and listing, next to a photograph, success and failure rates for all the procedures for which they are responsible (it is shortly to be amended to show success rates for the procedures they have actually carried out). Otherwise, mortality data is available for individual hospitals, but not routinely to the public through Patient Choice. It can, however, be found with persistence, and is sometimes published in the newspapers.

In Wales, it seems not to be available to the public at all. For more than a year, in conjunction with BBC Wales and the Centre for Health Economics at York, we have been trying to persuade various parts of the Welsh Health Service either to disclose how many people die in each hospital or to allow access to the hospital episode statistics to allow the calculation to be made independently. The degree of resistance to disclosing this elementary piece of information is baffling and illuminating. The Welsh Health Service argues that the data might compromise patient confidentiality, but refuses to produce even the total mortality figures for the whole country, from which the risk of any individual patient being identified is nil. So we do not know how well the whole system has been performing in even this modest respect, let alone individual hospitals. It is quite true that the data would need interpreting with care because of the high likelihood that some unique local circumstances will affect some local mortality rates, but this is not a sufficient excuse for making a state secret of them. In England, academics and the media have had access to this kind of information for twelve years. Patients in England do not seem to have suffered gross abuses of their confidentiality as a result. The Welsh authorities tell us that they are now beginning to do their own analysis of the data. Whether the public will be allowed to see it is another matter.

Not that this data would get us all that far: "It's all right, you'll live" is a poor measure of most treatments with not much relevance, one hopes, to a hip transplant, for example. Most people want a better guide to the quality of care they will receive than whether they are likely to survive it, but it would at least be a start and might, should there be serious problems, at least alert us to them.

A culture that respected data, that put proper effort into collecting and interpreting statistical information with care and honesty and an understanding of its limitations, that valued statistics as an aid to understanding, and took pains to find out what was said by the numbers we have already got, that regarded them as something more than a political plaything, a culture like this would, in our view, be the most valuable improvement to the conduct of government and setting of policy Britain could achieve.

What can we do on our own?

There are times when we are all whistling in the dark. And sometimes, in a modest way, it works: you know more than you think you do. We have talked about cutting numbers down to size by making them personal, and checking that they seem to make human sense. A similar approach works when you need to know a number, and feel you haven't a clue.

Here is one unlikely example we have used with live audiences around the UK and on BBC Radio. How many gas stations are there in the UK? You can ask the same question of the United States.

Not many people know the answer, and the temptation is to feel stumped. But making the number personal can get us remarkably close. Think of the area you live in, and in particular of an area where you know the population. For most of us that is the town or city we live in. Now think about how many gas stations there are in that area. It is hard if you have only just moved in, but for most adults this is a fairly straightforward task, and people seem very good at it. Now divide the population by the number of gas stations. This gives you the number of people per gas station in your area. For us in the United Kingdom, the answer was about 1 gas station for every 10,000 people. Most British people we have asked give answers that lie between 1 for every 5,000 and 1 for every 15,000.

We know the total population of the United States is about 300,000,000 but also that the country is much larger and less densely populated than the United Kingdom, so will probably have more gas stations per member of the population. So we just need to divide the population by the number of people we estimate for each gas station. With one gas station for every 5,000 people, the answer is 60,000 gas stations. With one per 2,500, the answer is 120,000 gas stations. The correct answer is about 120,000 (121,466 according to the 2002 Economic Census). The important point is that almost everyone, just by breaking things down like this, can get an answer that is roughly right. Using the same ideas would produce roughly accurate numbers for how many schools there are, or hospitals, or doctors, or dentists, or out-of-town shopping malls.

All that is happening is that rather than being beaten by not knowing the precise answer, we can use the information we do have to get to an answer that is roughly right, which is often all we need. As long as we know something that is relevant to the question, we should be able to have a stab at an answer. Only if the question asks about something where we have absolutely no relevant experience will we be completely stumped.

The best example of this we could come up with was "How many penguins are there in Antarctica?" Here, it really did seem that unless you knew, you could bring very little that helped to bear on the question. Apart from penguins, you will be surprised by how much you know.

SHOCK FIGURES 10

WAYWARD TEE SHOTS

News is filled with the cult of the shock figure, numbers that demand amazement or alarm. A number comes along that looks bad, awe-inspiringly bad, much worse than we had thought; big, too, bigger than guessed; or radically different from all we thought we knew.

Resist. When a number appears that's out of line with others, it tells us one of three things: (a) this is an amazing story, (b) the number is wrong, (c) it has been misinterpreted. Two out of three mean the story is wasting your time, because the easiest way to say something shocking with figures is to be wrong. Outliers—numbers that don't fit the mold—need especial caution: their claims are large, the stakes are high, and so the proper reaction is neither blanket skepticism, nor slack-jawed credulousness, but demand for a higher standard of proof.

Greenhouse gases could cause global temperatures to rise by more than double the maximum warming so far considered likely by the Intergovernmental Panel on Climate Change (IPCC), according to results from the world's largest climate prediction experiment, published in the journal *Nature* this week.

These were the words of the press release that led to alarmist headlines in 2005. It continued:

The first results from climateprediction.net, a global experiment using computing time donated by the general public, show that average temperatures could eventually rise by up to 11°C [19.8°F], even if carbon dioxide levels in the atmosphere are limited to twice those found before the industrial revolution. Such levels are expected to be reached around the middle of this century unless deep cuts are made in greenhouse gas emissions.

Chief Scientist for climateprediction.net David Stainforth, from Oxford University, said: "Our experiment shows that increased levels of greenhouse gases could have a much greater impact on climate than previously thought."

There you have it: 11°C (19.8°F) and apocalypse. No other figure was mentioned.

The experiment was designed to show climate sensitivity to a doubling of atmospheric carbon dioxide. Of the 2,000 results, each based on slightly different assumptions, about 1,000 were close to or at 3°C (5.4°F). One result was 11°C (19.8°F). Some results showed a fall in future temperatures. Guess which result was reported.

A BBC colleague described what happened as akin to a golfing experiment: you see where 2,000 balls land, all hit slightly differently, and arrive at a sense of what is most likely or typical; except that climateprediction.net chose to publicize a shot that landed in the parking lot. Of course, it is possible. So are many things. It is possible your daughter might become pope, but we wouldn't bet on it, at least not until she made cardinal. As numbers go, this was the kind that screamed to be labeled an outlier, and to have a red-flag warning attached, not an excitable press release or front page headlines.

In January 2007, this time in association with the BBC, climateprediction.net ran a new series of numbers through various models and reported the results as follows: "The UK should expect a 4°C [7°F] rise in temperature by 2080 according to the most likely results of the experiment."

That is more like it: "The most likely results" still may be wrong, as all predictions may be wrong, but at least they represent the balance of the evidence from the experiment, not the most outlying part of it.

None of this, incidentally, implies comfort to climate-change deniers: 7°F, even 5°F, still would bring dramatic consequences, though the case arguably would be strengthened—if more modest—by using this overwhelmingly likely result (according to this experiment), rather than one that could be dismissed as scaremongering. It is always worth asking if the number we are presented with is a realistic possibility, or the papal one.

"It could be you," say adverts for Britain's National Lottery, which is true, though we all know what kind of truth it is. In any case, responsible news reporting owes us something better than a lottery of possibilities. And yet there is an inbuilt bias to news reporting that actually runs in favor of outliers. "What's the best line?" every news editor asks, and every journalist knows that the bigger or more alarming the number, the more welcomed it will be by the boss. In consequence, the less likely it is that a result is true, the more likely it is to be the result that is reported. If that makes you wonder what kind of business news organizations think they are in, often the answer is the one that interests and excites readers and viewers most. It seems that consumers of news, no less than producers, like extreme possibilities more than likely ones. No wonder we are so bad at probabilities. "Your daughter could be Pope," says the number in the headline on the newsstand. "Gosh, really?" say readers, reaching for the price of a copy.

This is a question to ask continually of surprising numbers: is this new and different, or does the very fact that it is new and different invite caution? Do the numbers mark a paradigm shift or a rogue result? The answer in the climate-change example is, we think, clear-cut. Even some of those involved in the research came to regret the prominence they gave to a freak number.

For a trickier—and more colorful—case that brings out the judgment between oddity and novelty, try the Hobbit.

Some 18,000 years old, with the consistency of thick mashed potatoes or blotting paper, they turned up in a soggy cave (described by the journal *Nature* as "a kind of lost world") and became a worldwide sensation when news of the find was reported in March 2003. The skeletal remains were soon nicknamed Hobbit Man, though the most complete of the skeletons might have been—it is still disputed—a woman of about thirty years old, given the name Little Lady of Flores, or Flo. They were discovered in the Liang Bua cave on the island of Flores in Indonesia; hence the scientific name for what was hailed as an entirely new species of human: *Homo floresiensis.*

At about 39 inches (1 meter) tall, the "Hobbits" were shorter than the average adult height of even the smallest modern humans, such as African pygmies (pygmies are defined as having average male adult height of 4.9 feet or 1.5 meters or less), and it was this that grabbed the popular imagination.

They also reportedly had strikingly long arms and a small brain. The joint Australian–Indonesian team of paleoanthropologists had been looking for evidence of the original human migration from Asia to Australia, not for a new species. But these little people seemed to have survived until more recently than any subspecies of human apart from our own and may, some believe, be consistent with sightings of little people known locally as Ebu Gogo, reported until the nineteenth century. Some wonder if they live on in an isolated jungle in Indonesia.

They certainly were extremely unusual, apparently like no other discovered remains of any branch of the genus *Homo:* not *Homo erectus,* nor modern-day *Homo sapiens,* nor the Neanderthals. But were they a new species?

Two reasons to pause before judgment came together on February 10, 1863 in New York City—at a wedding: Charles Stratton married Lavinia Warren in a ceremony attended, as reports of the time captured that famous day, by the "haut ton" of society. At the time, Stratton was thirty-three inches tall, Warren thirty-four inches. A few inches shorter even than the Hobbits, but intellectually unimpaired and leading a full life, they had a fine, custom-built home paid for by their many admirers and Stratton's wealth from years as a traveling exhibit for P.T. Barnum, for whom he appeared under the stage name General Tom Thumb.

The usually staid *New York Observer* reported the wedding as the event of the century, if not unparalleled in history:

> We know of no instance of the kind before where such diminutive and yet perfect specimens of humanity have been joined in wedlock. Sacred as was the place, and as should be the occasion, it was difficult to repress a smile when the Rev. Mr. Willey, of Bridgeport, said, in the ceremony—"You take this woman," and "You take this man" &c.

Stratton and Warren obviously were the same species as the rest of us, perfect specimens according to the *Observer.* Their four parents and nine of their brothers and sisters were a more typical height, though the last of Lavinia's sisters, Minnie, was shorter even than she. Their existence within one species tells us, if we didn't know it already, that the human genetic program can produce extreme variation. The world's smallest man on record was Che Mah at two feet two inches. Robert Wadlow, the world's tallest, was eight feet 11 inches, more than four times as tall.

So imagine that our newlyweds had set up home on an island in Indonesia, started a family, and that their remains went undiscovered until the twenty-first century. How might we describe them? Would we recognize them for what they were, outliers in the wonderfully wide spectrum of human variation, or wonder perhaps if they, too, were a separate species?

The argument about whether Hobbit Man was a different kind, or simply an extreme variation of an already known kind, still rages. Most recently it has been argued that it was indeed a new species, following scans of the remains of the skull and a computer-generated image of the shape of the brain by a team at Florida State University.

The team was trying to determine whether the cave dwellers from Flores could be explained as the result of disease (microencephaly—having an exceptionally small brain—is a condition we know today), though there may have been some other, unknown syndrome that caused their particular constellation of physical characteristics. Professional rivalries and fights over access to the remains by eager researchers have complicated the investigation. "We have no doubt it is not a microcephalic," said Dean Falk, a paleoanthropologist that worked with the team who discovered the remains. "It doesn't look like a pygmy either." The journal *Nature*, which first published news of the Hobbit discovery, headlined the latest research: "Critics silenced by scans of Hobbit skull."

That assessment proved optimistic. Robert Eckhardt of Penn State University and a skeptic, reportedly said he was not convinced:

"We have some comprehensive analyses underway that I really think will resolve this question. The specimen has multiple phenomena that I would characterize as very strange oddities and probably pathologies." In 2008, Dr. Peter Obendorf from the School of Applied Science at RMIT University, Melbourne, and colleagues, also said they believed the little people were not a new species, but had developed a dwarfism condition because of severe nutritional deficiencies.

Fortunately, it is not necessary to settle the argument to make the point: unexpected and extreme findings might tell us something new and extraordinary; they might also be curious, but irrelevant to the bigger picture. The fact that they are extreme and unexpected is as much a reason for caution as excitement.

Imagine that the heights of every adult human in the UK were plotted on a graph. Most would bunch within four inches or so of the average for men of about five feet ten inches and five feet four inches for women (Health Survey for England 2004). Some, but not so many, would be further along the graph in one direction or another, at which point we might begin to call them tall or short. A very few, including Robert Wadlow, Che Mah, Lavinia Warren, and Charles Stratton, would be at the far ends. The problem with the Hobbit is to decide whether it is simply a human of a kind we already know, from the extreme end of the graph, an outlier with an illness, or if it tells us that we need to create a new graph entirely, and change our assumptions about human evolution. The argument continues.

Remember our three possibilities when faced with a shock figure—an amazing story, a wrong number, or a misinterpretation—and then apply them to the discovery of an adult skeleton three feet tall identified as a new species. Of the three, which is most likely?

Is the Hobbit genuinely new, different, and warranting a change to the whole map of human evolution? Or is it, like Charles Stratton perhaps, one of *Homo sapiens*' very own, but slightly erratic tee shots?

If the number is a genuine Hobbit, fine, we will have to change our thinking. But strange as Tom Thumb is, well, we knew already that the results of human reproduction can be surprising from time to time, that there will always be outliers, in all sorts of ways, but they remain, quite obviously, human. Statistics teaches us that outliers are to be expected, and so there is nothing abnormal about them, but they are certainly atypical. If that's all that was discovered in Liang Bua cave—an atypical human—it doesn't tell us anything new at all; in the scatter plot of human life, it's one wayward dot, and might be a guide to nothing more than its unusual self.

In fact, outliers usually are far less interesting than Tom Thumb. The real Charles Stratton fascinates; most statistical outliers on the other hand are not human but products of an experiment, survey, or calculation, observations that are (by definition) atypical, and one of the first principles of statistics is that suspicious data or outliers may need to be rejected, not least because there's every chance that the outlier is simply wrong—measured or recorded inaccurately. Tom Thumb at least was real, even if on the fringes of the distribution. Giving credence to statistical outliers from forecasts, on the other hand, indulges fantasy.

The point is that outliers can be a routine fluke of the system, an erratic moment, and there will always be such blips in any distribution of height, house prices, weather forecasts, or whatever. They need not be a cause for alarm. They are not necessarily a revelation.

If your business is catching drug cheats in sports, the natural variability that means some people will always be outliers is a headache. Many of the drugs that people take to make them faster or stronger are substances that are already in their system naturally. Athletes want a little more of what seems to make them good.

So cheats are detected by looking for those with extreme levels of these substances in their body. In essence, we identify the outliers and place them under suspicion. The hormone testosterone, for example, occurs naturally, and is typically found in the urine in the ratio of 1 part testosterone to 1 part epitestosterone, another hormone. The World Anti-Doping Agency says there are grounds for suspicion that people have taken extra testosterone in anyone found with a ratio of 4 parts testosterone to 1 part epitestosterone. The threshold used to be 6 to 1, but this was considered too lax.

There are two problems. First, there have been documented cases of outlying but entirely innocent and natural testosterone to epitestosterone (T/E) ratios of 10 or 11 to 1, well above the level at which the authorities become suspicious. Second, there are whole populations, notably in Asia, with a natural T/E ratio below 1 to 1, who can take illegal testosterone with less danger of breaching the 4 to 1 limit. In short, there is abundant variation.

The first person to be thrown out of the Olympics for testosterone abuse was a Japanese fencer with an astronomical T/E ratio of about 11 to 1. The Japanese sporting authorities were so scandalized they put him in the hospital, under virtual arrest, where his diet and medication could be strictly controlled. They found that his T/E ratio remained unchanged; he was a natural outlier. It took some believing that such odd results could mean nothing, but the evidence was incontrovertible. According to Jim Ferstle, a journalist we interviewed who has followed the campaign against drug taking in sports for twenty years, the man never received so much as an apology.

Until the late 1990s this was the only test for testosterone abuse, even though we have little idea what proportion of the population would have been naturally above the 6 to 1 ratio that then applied, nor an accurate idea of how many would be naturally above the 4 to 1 ratio that applies now. To add to the problem, it is well known that alcohol can temporarily raise T/E ratios, particularly among women. So there is a danger that if the outliers are simply rounded up and called cheats, there will be honest people unjustly accused.

Fortunately, there is now a second test, which detects whether the testosterone in a suspect sample is endogenous (originating within the body) or exogenous (originating outside). This test, too, has failed in experiments to catch every known case of doping (a group of students was given high levels of testosterone by a Swiss sports-medicine clinic and then tested, but not all were identified by the test as having been doped). This second test has not yet been shown falsely to accuse the innocent, but it can produce an "inconclusive" result. One athlete, Gareth Turnbull, an Irish 1500-meter runner, told us that he had spent about 100,000 euros on lawyers defending himself against a charge of illicit testosterone use after an adverse T/E test result and an "inconclusive" second test for the source of the testosterone, all following a night's heavy drinking. He eventually was cleared in October 2006, and in 2008 agreed to a financial settlement with his national athletics authority, but not before, as he puts it, it became impossible to Google his name without coming across the word "drugs."

As with height and the Hobbit, we need to remember that abnormal is normal, there are always outliers, and we should fully expect to see the computer spit out a figure of 20°F or higher—sometimes; but we should also acknowledge that it may tell us nothing. And if we wish to slip in a change of definition at some point—wherever that point is—at which it is decreed that normal stops and suspiciously abnormal starts, in order to label everyone beyond that point a potential cheat, or a new species, we have to be very sure indeed that redefinition is warranted. If those outliers are produced by forecasts or computer simulations, we might want to discard them altogether, or at least qualify them with the thought of the pope taking a wayward tee shot.

Certain words invite suspicion that an outlier is in play. When we see phrases of the kind "may reach," "could be as high as," and "potentially affect," it's worth wondering whether this is the most likely, or the most extreme possibility (and therefore one of the most unlikely), and then to ask how far adrift it is from something more plausible. Outliers will always pop up from time to time, particularly in forecasts, but these forecasts seldom come to pass. As a mildly entertaining game—each time you see the words "could potentially affect" or similar, add the mental parentheses: "but probably won't."

COMPARISON 11

I f we compare thee favorably to a summer's day, you might accept it as a compliment, but not the basis for a ranking chart. People and weather are categorically different, obviously, and such comparisons impossible—with apologies to Shakespeare—without plenty of definitional fudge. In a sonnet, we approve, and call it metaphor; but in politics . . .

And yet politics loves comparison. It is the staple of argument. Every "this-is-better-than-that" attempts it. Lately that enthusiasm has spilled into the way we're encouraged to think of schools, hospitals, crime, and much else, through the device of ranking charts and their raw material, the performance assessment: how one compares with another, who's up, who's down, who's top, who's bottom, who's good, bad or failing, who shows us "best practice." Comparison has become the supreme language of government. It is now, in many ways, the crux of public policy, in the ubiquitous name of offering informed choice.

But politics has precisely that bad habit of fudging the people/weather problem, and of overlooking definitional differences. To detect this, the principle to keep in mind is one everyone already knows, but has grown stale with overuse. It is as true and relevant as ever, however disguised in ranking charts or performance indicators, and it is this: is the comparison of like with like?

Former presidential contender and mayor of New York Rudolph Giuliani has survived prostate cancer. In August 2007, as the campaign for the presidential primaries kicked off, he used that experience in a New Hampshire radio advertisement to make this dazzling political comparison:

"I had prostate cancer, five, six years ago," he said. "My chances of surviving prostate cancer—and thank God I was cured of it—in the United States, 82 percent. My chances of surviving prostate cancer in England, only 44 percent under socialized medicine."

Devastating, if true. A health-care system often compared favorably to America's was in fact only half as good at curing cancer. There was even a (somewhat-dated) bar graph from the Commonwealth Fund, a think tank, which seemed to support the argument: The UK's National Health Service, relatively speaking, was a killer.

Was Rudy right? Is the comparison fair? The Commonwealth Fund disowned his interpretation of its data and a simple question shows why: is it likely that prostate cancer is 2.8 times as common in the United States as in Britain—some protective elixir in warm beer perhaps? That is what these figures suggest: incidence of 136 cases for every 100,000 men in the United States, but only 49 cases per 100,000 men in the UK. If he is right, we have to explain why American men are falling to this illness dramatically more often

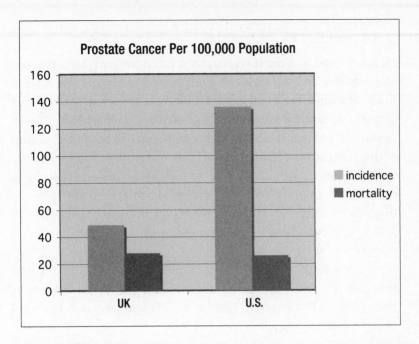

Prostate Cancer Per 100,000 Population

than men from any other developed country in the world for which we have data.

That sounds unlikely. Is there a more plausible explanation? One simple possibility is that having an illness and having it diagnosed are different (not everyone finds their way to the doctor at the same point). Perhaps it's not that three times as many have the illness in the United States, but that nearly three times as many are diagnosed.

It's among insidious differences that comparison comes unstuck. Rudy's comparison was ingenious but inumerate. Bump up the number of diagnoses, when deaths remain about the same, and presto, there's your massively higher "survival" rate. This graph says only a little about the effectiveness of the treatment for prostate cancer in the two countries; it hints at much more about the trend in the United States to early diagnosis.

In fact, the United States does have genuine reason to feel satisfied, beating the UK on most international comparisons of cancer treatment, so far as those comparisons can be trusted. Even this one suggests that fewer people die from prostate cancer in the United States than in Britain: 26 per 100,000 men, compared with 28 per 100,000. Not twice as good, as Rudy claimed, nor anything like it, but better, and that result might owe something to those higher rates of diagnosis and the fashion in the United States for health screening from a younger age. Though it might also be because survival is defined as living for five years beyond diagnosis. So if people are diagnosed earlier, they probably have more years left to live anyway and so appear to have a better survival rate even if the doctors do nothing.

A convoluted argument, you probably would agree. The ifs and buts pile up, different cultural practices raise unanswerable questions. But that's the point. Comparison is seldom straightforward once you start to dig a little.

All in all, since there's some evidence to suggest that Americans really do seem to have more cancer than others, even after adjusting for the propensity to early diagnosis, the American mortality rate might well be better than most other countries, but probably only a little better. Though, incidentally, diagnosis is not always a blessing. If it leads to treatment, the side effects can include infertility, impotence, and incontinence. Since more people die *with* prostate cancer than *from* it, doing nothing will in many cases do no harm, and might even prevent some.

"Eight out of ten survive," "four out of five prefer," "one in four this," "99 percent that" . . . all apparently simple forms of counting turned into a comparison by words to the effect . . . "unlike over there, where only 70 percent . . ." etc.

But "eight out of ten" what? Out of all those who have prostate cancer, or only those whose cancer comes to the attention of a doctor? Rudy's comparison fails because he picks his survivors from different groups, the more frequently diagnosed U.S. group with the less frequently diagnosed UK group. It's a tasty little ploy, though whether in this case accidental or deliberate, who knows?

With comparison, all counting's definitional snags multiply massively, since we define afresh with every comparison we make. To repeat the well-known essence of the problem: are we defining both groups in the same way? Are we comparing like with like in all respects that matter?

In the last fifteen years or so, the UK has seen a massive expansion of comparison through ranking charts, performance indicators, and the like, an explosion of standard setting unparalleled in British administrative history. Against these standards, just about everyone in the public (and some in the private) sector now seems to be judged. There are hints that the United States—with the emphasis, for example, on comparisons of performance against a benchmark in the No Child Left Behind education policy—might be moving the same way. The shock for the UK—and a warning for the United States—has been discovering how the categories for comparison seem to multiply. Things just won't lie down and be counted under what politicians hoped would be one heading, but turn out to be complicated, manifold, infernally out of kilter. Schools turn out never to be quite the same thing, given their different intakes, locations, and sizes. Hospitals always find reasons why their experience can't be compared with that of their neighbors. This is not only special pleading; some of the complaint is legitimate. If one hospital was built at the time of Queen Victoria, and another opened last year, even their different heating bills cause arguments. Counting in such circumstances is prone to an overwhelming doubt: what is really being counted?

The comparison of schools, hospitals, police forces, local councils, or any of the multitudes of ranked and performance assessed first ought to be an equal race. But it rarely is, and seldom can be.

Life is messier than that, the differences always greater than anticipated, and so often so much more than just a detail. So we have to decide, before ignoring these differences, if we are happy with the rough justice that implies. The exercise still might be worth it, for all sorts of reasons, but before making the call it pays to understand the trade-offs.

A prime example is the immense effort on both sides of the Atlantic to tell parents how their local schools compare. As an ambition for accountability, it sounds laudable and simple. But whereas the UK government began by treating all schools as much the same thing when it published results that were quickly and inevitably put into ranking charts, today some of those ranking charts (there's more than one version) include an elaborate and, for most parents, impenetrable calculation that tries to adjust the results for every state school in the country according to the characteristics of the pupils in it. If the local children are poor, they tend to do less well, and not necessarily because the school is bad. So the school results are now weighted according to deprivation, and a host of other, sometimes surprising characteristics. Although the comparison started by claiming to be a test of merit, it has collapsed into an unending spat about underlying differences.

The standardized test results that have developed in the United States in the last twenty years—given further momentum by the Bush administration's No Child Left Behind education policy—appear to have much in common with the course of education policy in the UK. Viewed from the UK, it is striking how the debate about the renewal of NCLB in 2008 has mirrored earlier rows about British education policy. Given that the stakes in the United States are arguably even higher than in the UK—failure to achieve proficient standards can lead to financial penalties and school closure—we offer the UK story as a cautionary tale to American readers.

As soon as it was decided to administer standardized tests to all UK pupils at several points in their school lives, the government foresaw that demand for publication of the results would be over-whelming. So the results were published, school by school, across the country, from 1992. Did the government expect to still be making fundamental revisions to the performance tables fifteen years later? Almost certainly not, but in 2007 performance tables for schools faced their third fundamental reform, turning upside down the ranking—and the apparent quality of education offered—of some schools. Without any significant change in their examination results, many among the good appeared to become poor overnight, and, among those formerly struggling, some suddenly were judged to excel. Out went the old system of measurement, in came the new. The public, which for several years had been told one thing was abruptly told, in some cases, the opposite. The government called it a refinement.

This history of school comparison in Britain is a fifteen-year lesson in the pitiless complexity of making an apparently obvious measurement in the service of what seemed a simple political ambi-tion: let's tell parents how their local schools compare. At least "simple" is how it struck most politicians at the time. One conclu-sion could be that governments are also prone to failures to distin-guish abstraction from real life, still insisting that counting is child's play.

The first tables in 1992 were straightforward: every school was listed together with the number of its children who passed five GCSEs (the exam taken by almost all UK sixteen-year-olds at the end of their compulsory education) at grade C or above. Though this genuinely had the merit of simplicity, it was also soon apparent that schools with a more academically able intake achieved better results, and it wasn't at all clear what, if anything, a school's place in the tables owed to its teaching quality.

For those schools held up as shining examples of the best, this glitch perhaps mattered little. For those fingered as the worst, par-ticularly those with high numbers of special-needs pupils or pupils whose first language was not English, it felt like being condemned by official stupidity, and was maddening. As U.S. commentators have also recognized, measuring what a child knows at any one time is not the same as measuring what that child has learned in any particular school.

What's more, the results for any one school moved from year to year, often with pronounced effects on the school's ranking-chart position. Professor Harvey Goldstein from Bristol University, an expert in ranking charts, told us: "You cannot be very precise about where a school is in a ranking. Because you only have a relatively small number of pupils in any one year that you are using to judge a school, there is a large measure of uncertainty—what we call an uncertainty interval—surrounding any numerical estimate you might give. And it turns out that those intervals are extraordinarily large; so large, in fact, that about two thirds to three quarters of all secondary schools, if you are judging by GCSE (age sixteen) or A-level (age eighteen) results, cannot be separated from the overall national average. In other words there's nothing really that you can say about whether the school is above or below the average of all pupils."

So the charts were comparing schools that were often unalike in kind, then straining the data to identify differences that might not exist. They were counting naively and comparing the tally careless as to what they were really counting. Some schools, conscious of the effect of the tables on their reputation—whether deserved or not—began playing the system, picking what they considered easier subjects, avoiding mathematics and English, even avoiding pupils—if they could—who they feared might fail, and concentrating on those who were borderline candidates, while neglecting the weakest and strongest for whom effort produced little reward in the rankings.

So the comparison, which had by now been the centerpiece of two governments' education policies, was revised, to show how much pupils in each school had improved against a benchmark of performance when they were aged eleven. Once again, the parallels with the U.S. experience—where a great deal of debate in 2008 has been about finding a way to measure how much students gain during the year—are plain to see.

This was an attempt in the UK to measure how much value the school added to whatever talent the pupil arrived with, not year by year, as discussed in the United States, but much more broadly—by comparing performance at two or three key stages of development in the child's school life. But these so-called value-added tables were nothing of the kind and unworthy of the name. (David Blunkett, who was education secretary at the time, described them to us as "unsatisfactory." When the minister responsible is dissatisfied, you can be sure there is a problem.) The benchmark used at age eleven was an average of all pupils at each grade. Many selective schools were able to cream off the pupils above the average of their grade and appear, once those pupils reached sixteen and were measured again, to have added a huge amount of value to them. In fact, the above-average value in these children's test results had been there from the start. These tables, misleading and misnamed, were published for four years.

Then another revision was announced, this time to require school results to include mathematics and English in the subjects taken at GCSE, a change prompted partly by the suspicion that some schools were improving their results, if not their standards, by avoiding "hard" subjects. In one case, this change caused a school in east London to slump from 80 percent of pupils achieving five passes at GSCE grade C, to a success rate of 26 percent.

Then there was a third major revision, known as contextual value added (CVA), which admitted the weaknesses of ordinary value added and aimed to address them by making allowance for all sorts of factors outside the school's control that are thought to lower performance—factors such as coming from a poorer background, a first language other than English, having special needs, being a boy, and half a dozen others. CVA also set pupils' performance against a more accurate benchmark of their own earlier ability.

In 2006, prior to the full introduction of CVA in early 2007, a sample of schools was put through the new calculation. What did the change in counting do to their position in the tables? One school, Kesteven and Grantham Girls' School, went from 30th in the raw GCSE tables to 317th out of the 370 sampled. Another, St. Albans C of E School in Birmingham, traveled in the opposite direction from 344th to 16th. Parents could be forgiven for wondering, in light of all this, what the comparisons of the past fifteen years that had put so many millions of them into a rabid panic about school choice had actually told them.

And there, so far, ends the history, but not the controversy. The CVA tables—complicated and loaded with judgments—have moved far from the early ideal of transparent accountability. It also turns out that the confidence intervals (how big the range of possible ranking-chart positions for any school must be before we are 95 percent sure that the correct one is in there) are still so large that we cannot really tell most of the schools apart, even though they will move around with much drama from one year to the next in the published charts. And in thinking about value added, it has dawned on (nearly) everyone that most schools are good at adding different kinds of value—some for girls, some for boys; some for high achievers, some for low; some in physics, some in English—but that the single number produced for each school can only be an average of all those differences. Few parents, however, have a child who is so average as to be 50 percent boy and 50 percent girl.

One significant repair looks like an accident; three wholesale reconstructions in fifteen years and you would want your money back. Unless, of course, a fair comparison was not really what they were after, but rather a simple signal to say which schools already had the most able children.

Some head teachers do report great benefits from the emphasis on performance measurement brought about by ranking charts, particularly with the new concentration on measures of value added. They have felt encouraged to gather and study data about their pupils and use this to motivate and discuss with them how they might improve. They pay more attention, they say, to the individual's progress, and value the whole exercise highly. That must be welcome and good.

And it would be absurd to argue *against* data. But it is one thing to measure, quite another to wrench the numbers to a false conclusion. Ministers often said that ranking charts should not be the only source of information about a school, but it is not clear in what sense they contributed *anything* to a fair comparison of school performance or teaching quality. Make a comparison blithely, too certain of its legitimacy, and we turn information into a lottery. As Einstein is often quoted as saying, "information is not knowledge."

Why was the UK story so hapless? In essence, for one reason: overconfidence about how easy it is to count and compare. There is much in life that is only sort of true, but numbers don't easily acknowledge sort ofs. They are fixed and uncompromising, or at least are used that way. Never lose sight of the coarse compromise we make with life when trying to count it.

In the United States, the comparison is further complicated by the fact that states are free to choose different cutoff points for proficiency. So one state might say that 90 percent of its children meet the required standard, while its neighbor can claim only 30 percent, but the lower score might actually hide a higher standard. Make the test easy enough and anyone and everyone can pass, but it doesn't mean the education is any good.

Perilous as comparison is within a single country, it pales besides international comparison. With definitions across frontiers, we really enter the swamp. Not that we would know it, from the way they are reported.

For a glimpse of what goes wrong, let us say we're comparing levels of sporting prowess; that's one thing, isn't it? And let us agree that scoring twenty touchdowns in a season of American football shows real sporting prowess. And so let us conclude that Zinedine Zidane, three times voted world soccer player of the year, having failed ever to score a single touchdown in American football, is lousy at sports. The form of the argument is absurd, and also routine in international comparison.

Whose health system is better, whose education? Who has the best governance, the fewest prison escapes? Each time things are measured and compared on the same scale, it is insisted that in an important way they are the same thing; they have a health system, we have a health system, theirs is worse. They teach math, we teach math, but look how much better their results are. They have prisons, we have prisons, and on it goes.

Visiting Finland, Christopher Pollitt of Erasmus University in the Netherlands was surprised to discover that official records showed a category of prisons where no one ever escaped, year after year. Was this the most exceptional and effective standard of prison security? "How on earth do you manage to have zero escapes every year?" he asked a Finnish civil servant.

"Simple," said the official, "these are open prisons."

Britain experienced a moral panic in early 2006 at the rate at which inmates were found to be strolling out of open prisons as if for a weekend ramble. By comparison, this seemed a truly astonishing performance. What was the Finnish secret?

"Open prisons? You never have anyone escape from an open prison?"

"Oh not at all! But because they are open prisons, we don't call it escape, we classify it as absent without leave."

It is Christopher Pollitt's favorite international comparison, he says. When you reach down into the detail, he argues, there are hundreds of such glitches. Finland did not, as it happens, boast the most effective prison security in the world, nor as some might have wanted to conclude from a comparison of the numbers of "escapes" alone, had it won, through heartwarming trust and a humane system, sublimely cooperative inmates.

At least we don't think so, although to be honest, we are not at all sure. Explanations are no more robust than the data they are built on. It is ludicrous, the time we devote to this, but we seem to seek explanations for differences between nations—why we are good and they're bad, or vice versa—when, if we cared to look, we would find reason to doubt whether the differences even existed in the terms described.

The directorate of education at the Organisation for Economic Co-operation and Development offers this example of the way that an attempt to set questions measuring international standards in science teaching can come to grief.

Question: how do you stop heat from escaping from a building?
Answer: put in more windows.
Is this answer correct?

❑   ❑
Correct   Incorrect

Obvious? It was judged incorrect in temperate countries, correct in very cold countries (where putting in "more windows" was assumed to mean more layers of glazing—triple-glazing is common), and a stupid question in very hot countries (why would you want to stop heat from escaping from a building?).

International rankings are proliferating. We can now read how the United States compares with other countries on quality of governance, business climate, health, education, transport, and innovation, to name a few, as well as more frivolous surveys like an international happiness index—"the world grump league," as one tabloid reported it. "Welcome," says Christopher Hood from Oxford University, who leads a research project into international comparisons, "to Ranking World." The number of international governance rankings, he says, has roughly doubled every decade since the 1960s.

Of course, you want to know how well the United States scores in these rankings; making such comparisons is irresistible, and even a well-informed skeptic like Christopher Hood enjoys reading them. We'll tell you the answer at the end of the chapter. First, some self-defense against the beguiling simplicity of Ranking World.

"With a header in the twenty-seventh minute followed by a second in first-half injury time, playmaker Zinedine Zidane sent shock waves through his Brazilian opponents from which they would never recover . . . The French fortress not only withstood a final pounding from Brazil but even slotted in another goal in the last minute."

The words are by FIFA, the world governing body of soccer, describing France's victory in the final of the World Cup in 1998, as only soccer fans could. Two years on, the Gallic maestros, as FIFA probably would put it, stunned the world again, taking top spot in the league of best health-care systems compiled by the World Health Organization.

The United States was ranked fiftieth, a poor showing for a rich country and humiliating, if you believed the WHO survey was to be taken seriously. And though the WHO is a highly respected international organization whose ranking charts are widely reported, many, particularly in the United States, did not. (Britain finished a disappointing eighteenth and didn't much like it either.)

The great advantage a soccer league has over health care is that in soccer there is broad agreement on how to compile it. Winning gets points, losing doesn't; little more need be said (give or take the odd barroom post-match inquest about goals wrongly disallowed, and other nightmare interventions by the ref). Being that easy, and with the results on television on Saturday afternoons, it is tempting to think this is what ranking charts are like: on your head, Zidane, ball in the back of the net, result, no problem.

But for rankings of national teams, even FIFA acknowledges the need for some judgment. For international games, each result is weighted according to eight factors: points are adjusted for teams that win against stronger rather than weaker opposition, for away games compared with home games, for the importance of the match (the World Cup counting most), for the number of goals scored and conceded. Gone is the simplicity of the domestic league. The world rankings are the result of a points system that takes all these factors and more into account, and when the tables are published, as they are quarterly, not everyone agrees that they're right. It is an example of the complexity of comparison—how good is one team, measured against another—in a case where the measurement is ostensibly easy.

Observing France's double triumph, in soccer and health, Andrew Street of York University and John Appleby of the King's Fund health think tank set out, tongue in cheek, to discover if there was a relationship between rankings of the best health-care systems and FIFA rankings of the best soccer teams.

And they found one. The better a country is at soccer, the better its health care. Did this mean the U.S. team manager was responsible for the nation's health, or that the health secretary should encourage family doctors to prescribe more soccer? Not exactly: the comparison was a piece of calculated mischief designed to show up the weaknesses of the WHO rankings, and the correlation was entirely spurious.

They made it work, they freely admitted, by ignoring anything that didn't help, playing around with adjustments for population or geography until they got the result they wanted. Their point was that any ranking system, but especially one concerned with something as complicated as health care, includes a range of factors that can easily be juggled to get a different answer.

Some of the factors taken into account in the compilation of the WHO survey are: life expectancy, infant mortality, years lived with disability, how well the system "fosters personal respect" by preserving dignity, confidentiality, and patient involvement in health-care choices, whether the system is "client orientated," how equally the burden of ill health falls on people's finances, and the efficiency of health-care spending (which involves an estimate of the best a system could do compared with what it actually achieves). Most people would say that most of these are important. But which is most important? And are there others, neglected here, that are more important?

This monstrous complexity, where each factor could be given different weight in the overall score, where much is estimated, and where it is easy to imagine the use of different factors altogether, means that we could if we wanted produce quite different rankings. So Street and Appleby decided to test the effect on the rankings of a change in assumptions. The WHO had claimed that its rankings were fairly stable under different assumptions. Street and Appleby found the contrary. Taking one of the trickier measures of a good health system, efficiency, they went back to the 1997 data used to calculate this, changed some specifications for what constituted efficiency and, depending which model they used, found that a variety of countries could finish top. They managed, for example, to move Malta from first to last of 191 countries. Oman ranged from 1st to 169th. France on this measure finished from 2nd to 160th, Japan from 1st to 103rd. The countries toward the bottom, however, tended to stay more or less where they were whatever the specification for efficiency.

They concluded, "The selection of the WHO dimensions of performance and the relative weights accorded to the dimensions are highly subjective, with WHO surveying various 'key informants' for their opinions. The data underpinning each dimension are of variable quality and it is particularly difficult to assess the objectivity with which the inequality measures were derived."

In short, what constitutes a good health-care system is in important ways a political judgment, not strictly a quantitative one. The United States does not run large parts of its health system privately in a spirit of perversity, knowing it to be a bad system compared with other countries. It does it this way because, by and large, it thinks it best (though the result of the 2008 election might change that). Others disagree; but to insist that the United States be ranked lower because of its choices is to sit in judgment not of its health-care system but its political values. And the differences are not just political. At the most elementary level, what we count varies from one country to another. So what Germany spends on spas apparently counts as public health expenditure. In the United States, it falls under an entirely different category.

Ted Marmor, a professor of political science at Yale University, has said of the international comparison of health systems: "Misdescription and superficiality are all too common. Unwarranted inferences, rhetorical distortion, and caricatures all show up too regularly in comparative health policy scholarship and debates."

It is tempting, once more, to give up on all comparisons as doomed by the infinite variety of local circumstances. But we can overdo the pessimism. The number of children per family, or the number of years in formal education, or even, in a pinch, household income, for example, are important measures of human development and we can record them just about accurately enough across most countries so that comparisons are easy and often informative. The virtue of these measures is that they are simple, counting one thing and only one thing, with little argument about definitions. Such comparisons, by and large, can be trusted to be reasonably informative, even if not absolutely accurate.

The more serious problems arise with what are known as composite indicators, such as the quality of a health system, which depend on bundling together a large number of different measures of what a health system does—how well your doctor treats you in the operating room, how long you wait, how good the treatment is in hospitals, how comfortable, accessible, expensive, and so on, and where some of what we call "good" will really mean what satisfies our political objectives. If one population wants abundant choice of treatment for patients, and another is not bothered about choice, thinks in fact that it is wasteful, which priority should be used to determine the better system?

What is important, for example, for children to learn in math? In one 2006 ranking, Germany was ahead of the UK, in another the UK was ahead of Germany. You would expect math scores, of all things, to be easily counted. Why the difference?

It arose because each test was of different kinds of mathematical ability. That tendency noted at the outset to assume that things compared must be the same implies, in this case, that the single heading "math" covers one indivisible subject. In fact, the British math student turns out to be quite good at practical applications of mathematical skill—for example to decide the ticket price for an event that will cover costs and have a reasonable chance of making a profit—while German students are better at traditional math such as fractions. Set two different tests with emphasis on one or the other but not both, and guess what happens? The reaction in Germany to their one bad performance (forget the good one) bordered panic. There was a period of soul-searching at the national failure and then a revision of the whole math curriculum.

Though the need to find like-for-like data makes comparison treacherous, there are many comparisons we make that seem to lack data altogether. The performance of the American and French economies is one example. To parody only slightly, the perception outside France seems to be that it is a country of titanic lunch breaks, an overmighty, work-shy public sector, farmers with one cow each, and the tendency to riot when anyone dares mention the word competition.

America by contrast, the land of turbocharged capitalism, roars ahead without vacations or sleep. And if you measure the American economic growth rate, it is, averaged over recent years, about 1 percent higher than in France—a big difference.

Look a little more closely, though, and it turns out that the American population has also been growing about 1 percent faster than that of France. So it is not that the Americans work with more dynamism, just in more and more numbers. When we look at the output of each worker per hour, it turns out that the French produce more than the Americans and have for many years—their lead here has been maintained (though it is quite true that they don't work nearly as many hours a year, on average). Even the French stock market has outperformed the American, where $1 invested thirty years ago was worth (at time of writing) about $36, while in France it was worth $72 (October 2006).

None of these numbers is conclusive. All could be qualified further, by noting the smaller initial size of the French stock market, for example, or high levels of French unemployment compared with the United States. We are not claiming that the French economy outperforms the American. Summary comparisons of complicated things are not possible with single numbers. When comparing such monstrously complex animals as entire economies, remember once again how hard it is to see the whole elephant.

Meaningful comparison is seldom found in single figures. Exceptions are when the figures apply to a single indicator, not a composite, when there's little dispute about the definitions, and where the data will be reasonably reliable. One such is child mortality. There is no debate about what a death is and we can define a child consistently. There will, in some countries, be difficulties collecting the data so that the figures will be approximate, as usual. But nevertheless we can effectively compare child mortality across the world, noting, for example, a rate in Singapore and Iceland of 3 children per 1,000 aged under five, and in Sierra Leone of 283 children per 1,000 (*The State of the World's Children*, UNICEF, 2006) and we can be justifiably horrified.

More complicated comparisons require a great deal more care. But if care is taken, they can be done. In Aylesbury prison in 1998 one group of prisoners was given a combination of nutritional supplements; another was given a placebo. Otherwise they carried on eating as normal. The group that received the genuine supplements showed a marked improvement in behavior. The researchers concluded that improved nutrition probably had made the difference. The results had significant implications for criminal justice and behavior in general, but seem to have been effectively ignored by the Home Office, which refused us an explanation for its unwillingness to support a follow-up trial and finally acquiesced to a new study only in 2008.

Yet this comparison had merit. Care was taken to make sure the two groups were as alike as possible so that the risk of there being some lurking difference, what is sometimes known as a confounding variable, was minimized. The selected prisoners were assigned to the two groups at random, without either researchers or subjects knowing who was receiving the real supplement and who was receiving the placebo until afterward, so that any expectations they might have had for the experiment would not be allowed to interfere with the result. This is what is known as a double-blind randomized placebo-controlled trial. Since the experiment took place in prison, the conditions could be carefully controlled.

A clear definition of how misbehavior was to be measured was determined at the outset, and it was also measured at different degrees of severity, not just one. A reasonable number of people took part, some 400 in all, so that fluke changes in one or two prisoners were unlikely to bias the overall result. And the final difference between the two groups was large, certainly large enough to say with some confidence that it was highly unlikely to be caused by chance.

This is statistics in all its sophistication, where numbers are treated with respect. The paradox is that the experiment had to be complicated in order to ensure that what was being measured was simple. They had to find a way of ruling out, as far as possible, anything else that might account for a change in behavior. With painstaking care for what numbers can and cannot do, a clear sense of how the ordinary ups and downs of life can twist what the results seem to show us, if we are not wise to them, and a narrow, well-defined question, the researchers might just have hit on something remarkable.

As the prisons overflow and the effectiveness of strategies against reoffending seems in universal doubt, often owing to a failure to measure their effects carefully, this strategy—cheap, potentially transformative, carefully measured—remains neglected. Isn't this a little odd? Of course, there is still a possibility that the results fell victim to a rogue, confounding factor or measurement error, but the process seems to have been responsible enough. Nine years on, when we first came across these results, the failure to pursue the findings, to try to replicate the experiment to test again whether the results were chance, was mystifying. Weak numbers and bogus numbers are hurled about with abandon in many comparisons. Here, where the numbers were strongly suggestive and responsibly used, they were ignored. The new trial will be fascinating.

Finally, as promised, the answer to the question of where the United States ranks internationally. Taking the more serious rankings together, it is, according to Christopher Hood's analysis of the various rankings then available, bottom of those OECD countries for which there was data; thirteenth out of thirteen, one behind France and then the UK, similarly miserable in eleventh. But you are no longer interested in such a flawed and Byzantine comparison. Are you?

CAUSATION 12

THINK TWICE

This causes that. Press the remote and the channel changes. Plant the seed and it grows. When the sun shines, it is warmer. Sex makes babies.

Human (and sometimes animal) ability to see how one thing leads to another is prodigious—thank goodness, since it is vital to survival.

But it also goes badly wrong. From applying it all the time, people acquire a headstrong tendency to see it everywhere, even where it isn't. We see how one thing goes with another—and quickly conclude that it causes the other, and never more so than when the numbers or measurements seem to agree.

This is the oldest fallacy in the book, that correlation proves causation, and also the most obdurate. And so it has been observed by smart researchers that overweight people live longer than thinner people, and therefore it was concluded that being overweight causes longer life. Does it? We will see.

How do we train the instinct that serves us so well most of the time for the occasions when it doesn't? Not by keeping it in check—it is genius at work—but by refusing to let it sleep. Do not rest on the first culprit, the explanation nearest at hand or most in mind. Do not, like Pavlov's dogs, credit the bell with bringing the food, feel your mouth begin to water, and leave it at that. Keep the instinct for causation restless in its search for explanations and it will serve you well.

Loud music causes acne. How else to explain the dermatological disaster area witnessed in everyone wired to headphones audible at fifty paces?

That is a cheap joke. There are many possible causes of acne, even in lovers of heavy metal, the likelier culprits being teenage hormones and diet. Correlation—the apparent link between two separate things—does not prove causation: just because two things seem to go together doesn't mean one brings about the other. This shouldn't need saying, but it does, hourly.

Get this wrong—mistake correlation for causation—and we flout one of the most elementary rules of statistics or logic. When we spot a fallacy of this kind lurking behind a claim, we cannot believe anyone could have fallen for it. That is, until tomorrow, when we miss precisely the same kind of fallacy and then see fit to say the claim is supported by compelling evidence. It is frighteningly easy to think in this way. Time and again someone measures a change in A, notes another in B, and declares one the mother of the other.

Try the logic of this argument: the temperature is up, the coastline is eroding; therefore global warming is eroding the coastline.

Or this: the temperature is up, a species of frog is dying out; therefore global warming is killing the frogs.

Or how about this: the temperature is up, there are more cases of malaria in the East African highlands; therefore global warming is causing more malaria in the East African highlands. QED.

Convinced by these news stories from respectable broadcasters and newspapers? You shouldn't be; they are all causation/correlation errors, made harder to spot by plausibility (at least to some). Plausibility is often part of the problem, encouraging us to skip more rigorous proof and allowing the causation instinct to settle too quickly: it sounds plausible, so it must be right. Right? Wrong.

In all these cases, environmental campaigners noted that as one measurement—average global temperature—moved, so did another: the position of the coastline, the number of frogs, cases of malaria. They put these facts together and confidently assumed they had added two and two to make four and produce what they called compelling evidence, but we might prefer to call them classic cases of logical hopscotch, fit for, if not derision, at least serious skepticism. All these claims have been vigorously and credibly challenged, as we shall see.

It is worth saying here and now that this chapter is no exercise in climate change denial. We need to beware of another fallacy, namely that because campaigners sometimes make false claims about the effects that therefore no effects exist. That doesn't follow either. We can note in passing that some critics of global warming have been equally guilty of spectacular numerical sophistry. The point is that even with strong cases, perhaps especially with strong cases of which people are devoutly convinced, they get carried away, and are too quick to let causation rest wherever it helps them.

So this is a guide to a certain variety of failed reasoning, but it is a frequent failure in hard cases, given great impetus by numbers. If we state that a rare frog is dying out as a result of global warming, it sounds OK, but lacks beef; it would be more powerful if we could throw in some measurements and say that researchers believe the past decade of record temperatures has led to a 60 percent decline in the population of red-spotted tree frogs.

Put aside, if you can, your own convictions, and follow us to the point: we are concerned here only with how to avoid mistaking correlation for a causal relationship. Learn how and, whatever side you are on, you can get closer to something more important than this kind of conviction: understanding.

It is a peculiar hazard, this tendency to confuse causation and correlation, which is (A) well known and well warned against, yet (B) simultaneously repeated ad nauseam, making it tempting to say that A causes B. It is also a hazard far more widespread than debate about climate, if more easily spotted, it must be said, in examples like these:

People with bigger hands have better reading ability; so we should introduce hand-stretching exercises in schools.

In Scandinavia, storks are more likely to be seen on the rooftops where larger families live. Therefore storks cause babies.

Less obvious here:

Children who come further down the birth order tend to do less well in school tests. Therefore birth order determines intelligence.

Downright controversial here:

People with multiple sclerosis have lesions in the brain; so if we stop the lesions, we can stop the disease (i.e., the lesions cause the problem).

And here:

Girls at single-sex schools do better than girls in mixed schools, therefore single-sex schools are better for girls.

What seems often to determine how easily we spot causation/correlation errors is how fast a better explanation comes to mind: thinking of decent alternatives slows conclusions and sows skepticism. Once again, imagination can take you far (though a thirst for more data also helps).

Good prompts to imagination are these straightforward questions: what else could be true of the group, the place, the numbers we are interested in? What other facets do they share, what else do we know that might help to explain the patterns we see? This is where the instinct for seeing causation can be put to good use, by stretching it further than the first answer at hand.

Where shall our imagination begin? With the most comical of our examples. The statement that hand size in children correlates with reading ability is true; but true because . . . ?

Because we tend to read better as we grow up, mostly through maturing intelligence and education, and as we grow older our hands grow bigger. Bigger hands are a correlate of better reading, but the cause lies elsewhere; no case, then, for a program of hand stretching in schools.

Next, storks and babies. This is harder, since the true explanation is less easy to guess, and there really are more storks on homes with larger families. But where does the causation truly lie? Perhaps because the house tends to be bigger as family size increases, and with more roof space . . .

In both cases there is a third factor that proves to be the genuine explanation: age in the first, house size in the second. That is one typical way for causation/correlation error to creep in. Two things change at the same time, but the reason lies in a third.

Now we begin to see how it works, what about the others, all of which have made the news?

Sufferers of multiple sclerosis have lesions in the brain. The more advanced the illness, the worse the lesions. But do the lesions cause the progressive disability characteristic of that illness? It is plausible—for many years it was thought true—and when a drug called beta interferon, which seemed to arrest the lesions, was discovered it was used in the fervent hope that it would slow the disease.

There was only one way to verify the hypothesis, and that was by studying patients over many years to see how fast the illness progressed, relative to the number of lesions and the use of beta interferon. The results, when finally produced in 2005, were bitterly depressing: patients who have taken beta interferon do have fewer lesions, but are no better on average than others who have not. The progressive worsening of other symptoms seems to continue at the same rate in both groups. The lesions were found to be an effect, not a cause, of multiple sclerosis, and beta interferon was, said researchers, to use a mordant analogy, no more than a bandage that failed to treat the cause.

Birth order and intelligence is also tricky. It is, once again, true that the further along the birth order you are, the worse you tend to do in IQ tests: first-borns really do perform best, second-borns next best, and so on, not every time, but more often than not, and there is a plausible explanation (watch out!) that goes like this. The more children a family has, the less parental attention each receives: the first has lots, the second maybe half as much, and so on. This is believable, but does that make it true?

Let's try the imagination test: what else could be true as you pass along the birth order? At third or fourth, or even sixth or seventh, what is plainly true is that we are now looking at a big family. What do we know about big families? One thing we know is that they tend to be of lower socioeconomic status. Poorer people tend to have more children, and we also know that the children of poorer families tend, for various reasons, to do less well. So the further down the birth order you are, the more likely you are to come from a poorer family; not always, of course, but is this true often enough to be the explanation for what happens on average?

The evidence is not conclusive, but the answer is "probably," since it also turns out, when looking at the birth order of children from the same family, that no one has found a significantly consistent pattern of performance; the last born within the same family is, as far as we know, just as likely to do best in an IQ test as the first.

The causation/correlation mistake here has been to try to explain what happens across many families (richer, smaller ones tend to do better; larger, poorer ones not so well) and claim that it applies to birth order within any one family. It looks plausible (that word again) and it seems commonly believed, but it probably is wrong.

Next, gender and school performance. It is true that girls attending single-sex schools do better academically than girls who don't. Does this prove causation, i.e., does it prove that it's single-sex education that produces the better examination results? (We'll put aside the question of what it does to their social education, which is too normative a concept to measure.) Is it, in short, the lack of boys that did it?

Again, we must use our imaginations to ask what else is true of girls in single-sex schools. We must be restless in the search for causation and not settle on the obvious correlation as our culprit. The first thing that's true is that they have relatively wealthy parents; most of these schools are fee-paying. And what do we now know from the previous example about socioeconomic status and academic performance? Wealthier families tend to have, for whatever reason, academically higher performing children. Second, single-sex schools are more often selective, so that there will be a tendency to take the more able girls to begin with. So it is not surprising that single-sex schools do better: they take more able girls than other schools and these girls are usually from wealthier families. We have established that they ought to do better for all sorts of reasons, and this without taking into account any consideration of the effect of single-sex teaching.

Spare a thought for the statistician asked to settle this question who has to find a way of distilling school results to rid them of the effect of socioeconomic background or pupil selection by ability so as to isolate the gender effect. As far as they have been able to do that, the balance of statistical opinion, once they have made these allowances, is that there is no difference.

There *is* evidence that girls tend to make slightly less inhibited choices of subject in a single-sex school, and it will almost certainly suit some pupils—which may be reasons enough for wanting your daughter to attend—but it cannot be expected to secure better exam results in general than the girls would achieve had they attended a mixed school that took pupils of similar ability.

Perhaps now, if the sensitivity of these subjects hasn't created so much hostility that we've lost our readers, we can turn to one of the most sensitive of all: climate change.

First, malaria in East Africa. It has been known for some time that malaria in highland areas is hindered by low temperatures, which inhibit the growth of the parasites in the mosquito. The Tear Fund was one of several charities to produce evidence of an increased incidence of malaria in the East African highlands and to attribute it to climate change.

There were anecdotes: of the man from the highlands who had become landless and was living in poverty because he was bitten, got malaria, couldn't sustain work on the land, lost the land and was forced to work in bonded labor.

But when researchers looked closely at the records, they found no support for the argument. One of them, Dr. Simon Hay, a zoologist from Oxford University, said of the records for that specific area in contrast to global averages, that: "The climate hasn't changed, therefore it can't be responsible for changes in malaria." His colleague David Rogers, a professor of ecology, said that some groups responded to this by accepting that there was no change in average climate but arguing that there had been a change in variability of the climate. That is an intelligent proposition, knowing, as we now do, that averages can conceal a lot of variation. So they looked—and found no significant change there either. The researchers concluded that an increase in drug resistance is a more likely explanation for the observed increase in malaria—in this instance. This was a case where there was not even a correlation at the local level but an assumed connection between what was happening to climate globally and disease locally.

Mary Douglas, an anthropologist, wrote that people are in the habit of blaming natural disasters on things they do not like. But the loose conjunction in the back of the mind of two things both labeled "bad" is not a sound basis for believing that one causes the other.

The so-called first victim of climate change was the South American golden toad. "It is likely," said one campaigner, "that the golden toad lives only in memory."

J. Alan Pounds, from the Golden Toad Laboratory for Conservation in Costa Rica, acknowledges that the toads have been badly affected by a disease called the chytrid fungus, but argues: "Disease is the bullet, climate change is the gun."

In fact, the fungus does not need high temperatures, and was deadly to the toads anywhere between 39°F and 73°F. Alan Pounds is not convinced: "We would not have proposed the hypothesis we did if there was not such a strong pattern," he said. It is quite likely, most scientists believe, that climate change will wipe out some species. It is not at all clear in this case that it has already done so.

Climate change is confidently expected to result in a rise in sea levels. Rising sea levels may well cause coastal erosion. The temptation is to observe coastal erosion and blame a climate-change effect on sea levels, as several TV news reports have done, accompanied by dramatic shots of houses teetering on clifftops.

But take a moment to note that even the environmental campaigning group Friends of the Earth is hesitant. Though deeply concerned about the future effects of climate change on coastal erosion, it says there has been erosion at the rate of about 3.3 feet a year for the last 400 years in parts of East England: "This erosion over the centuries is a result of natural processes and sea-level rise from land movements. However, in recent years the rate of erosion appears to have increased at some points along the coast. The causes are poorly understood, but in addition to natural processes and sea-level rise, the effects of hard coastal defenses are thought to play an important role. Ironically, our attempts to defend against sea-level rise may actually add to coastal erosion." When Friends of the Earth is cautious, reporters might also think twice.

Untangling climatic causation from correlation is fiendishly hard. And though climate change may raise sea levels and make coastal erosion seriously worse in the future, it is hard to claim it has made a difference yet. In fact, the rate of sea-level rise was faster in the first half of the twentieth century than in the second. (Though some argue that climate change has already affected the frequency and severity of storms and that these have caused coastal damage.)

When a tidbit of evidence seems to our taste, the temptation is to swallow it. For the noncommitted, too, this kind of deduction has an appeal—to laziness. It doesn't demand much thought, the nearest suspect saves time, the known villain might as well do. Even with intellectual rigor, mistakes happen, as with beta interferon and multiple sclerosis. During medical trials of new drugs, it used to be customary to record anything that happened to a patient taking an experimental drug and say the drug might have caused it: "side effects," they were called, as it was noted that someone had a headache or a runny nose and thereafter this "side effect" was printed forever on the side of the packet. Nowadays these are referred to as "adverse events," making it clear that the cause was unclear and they might have had nothing to do with the medication.

Restlessness for the true cause is a constructive habit, an insurance against gullibility. And though correlation does not prove causation, it is often a good hint, but a hint to start asking questions, not to settle for easy answers.

There is one caveat. Here and there you will come across a tendency to dismiss almost all statistical findings as correlation-causation fallacy, a rhetorical cudgel, as one careful critic put it, to avoid believing *any* evidence. But we need to distinguish between casual associations often made for political ends and proper statistical studies. The latter come to their conclusions by trying to eliminate all the other possible causes through careful control of any trial, sample, or experiment, making sure if they can that there is no bias, that samples are random when possible. The proper response is not to trash every statistical relationship but to distinguish between those that have taken some thought and those that were a knee-jerk.

So what, finally, about the correlation at the beginning of this chapter, of being overweight and longevity? It is true that the data from the United States shows overweight people (overweight, not obese) living a little longer than thin people. So what other factors makes it hard to be sure if there is a direct causal link between putting on weight and adding years? One possibility is illness. Very ill people tend to be very thin. Put their fates into the mix and it has an effect on the results.

This argument is far from settled. A technical quarrel? Yes, but more importantly an imaginative and a human one. You don't need a course in statistics to be struck by the realization that sick people often get thinner. We can all make causation errors, and we are all capable of detecting them, if we think twice.

LAST WORD

Although this book has been about how numbers are used, it is best thought of as being not about numbers at all, but about life. That is another way of saying that what matters most is the practical point of it all, which is neither primarily to entertain, nor to ridicule, but to change. And the change we hope for is not only in people's enjoyment of, or facility with, numbers for their own sake. It is change in the way we understand experience, change to the way we are governed, to how we are informed, to the way we see and think. Numbers, at least the kind here, are how we often express hope and anxiety, the way we capture our well-being, often the way we assert political and moral values. The tiger might not be real, but the purpose is as real as they come.

Finally, to further the aim of simplicity, here—almost seriously— is simplicity simplified: a summary guide to seeing through the world of numbers, on one page.

Size is for sharing
Numbers are tidy; life isn't
People count (with backache)
Chance lurks
Stripes aren't tigers
Up and down happens
"Average = middle" = muddle
You can't see wholes through keyholes
Risk = people
Most counting isn't
No data, no story
Easy shocks are easily wrong
Thou art not a summer's day, sorry
This causes that—maybe

ACKNOWLEDGMENTS

The two authors named on the cover of this book stand at the end of a very long line of helpful, thoughtful, and encouraging people who have contributed either directly to the ideas in it or its production, or been part of a sharp-eyed community of numerate analysts of public affairs without which the whole enterprise could not even have begun. We have in some ways simply been the lucky ones who got to write it when its time had come. From the initial radio program that gave us so much impetus, to the book itself, and now for our continuing ambitions to change the culture of numbers in public argument, we owe a great many people a great deal of gratitude.

Helen Boaden was the far-sighted BBC Radio 4 controller who took the plunge; Nicola Meyrick the program's incisive editor since its inception. Mark Damazer was Helen's successor and became, to our delight, *More or Less* cheerleader in chief. We owe them huge thanks.

We have been lucky to work with some talented journalists who, one by one, came through the program as if it were a revolving door, bringing energy and a smile and leaving it, smile intact, to spread the word. Some of their reports were the basis for examples used here. Thanks to Jo Glanville, Anna Raphael, Ben Crighton, Adam Rosser, Ingrid Hassler, Sam McAlister, Mayo Ogunlabi, Jim Frank, Ruth Alexander, Paul O'Keeffe, Richard Vadon, Zillah Watson, our PAs Bernie Jeffers and Pecia Woods, and especially to Innes Bowen, whose tireless intelligence has been invaluable. Many others in Radio Current Affairs and Radio 4, studio managers, Internet and unit support staff, provided creativity and quiet professionalism that enabled us to spend our time as we should, pulling out our hair over the content. Thanks, too, to Gwyn Williams, Andrew Caspari, Hugh Levinson, and the many colleagues and kind reviewers, the hundreds who have written to us, the hundreds of thousands who have listened, all of whom, in one way or another, have egged us on.

There have now also been hundreds of interviewees and other direct contributors to the program, and hence to our thoughts in this book. To single out any one from the ranks of wise and willing would be unfair. All deserve our sincere thanks.

We would particularly like to acknowledge the extensive help, comment, and advice of Kevin McConway, the best kind of scrupulous and generous critic, and others at the Open University, and also Helen Joyce, Michael Ranney, Brad Radu, Mark Liberman (for the blog that drew our attention to the discussion of size in genetics), Rob Eastaway, Rachel Thomas, Gwyn Bevan, Richard Hamblin, and the assistance of Catherine Barton.

Andrew Franklin at Profile Books, razor sharp as ever, and his skillful colleagues, Ruth, Penny, Trevor, and others, somehow make publishing fun and humane, even while laboring against our awkwardness. Thanks, again. Erin Moore of Penguin Books in the United States got the point faster than anyone, and was a great source of encouragement and guidance on the preparation of the U.S. edition.

Finally, thanks to Catherine, Katey, Cait, Rosie, and Julia for their love, thoughts, support, and, what's most important, for being there.

We have done our best to avoid mistakes, but we know that we have failed to catch them all. We are very grateful to readers who spotted mistakes in the first edition and very much look forward to hearing from readers who spot more. Thank you in advance.

FURTHER READING

There's a growing library of popular science books on all that's fascinating about numbers—from stories about enduring mathematical theories to histories of zero and much more—and they're often intriguing and entertaining. This list, by contrast, is of books about how to make sense of the sort of numbers that find their way into the news or are likely to confront us in everyday life. Books, in other words, to further the aim of this one.

The best of those on bad statistics is still Darrell Huff's classic, first published in 1954 and now available as *How to Lie with Statistics* (W. W. Norton, 1993). It is short, lively, and timeless, and gives numerous amusing examples. If you were inclined to think people must have grown out of this kind of rudimentary mischief by now, you'd be wrong.

Joel Best comes at similar material as a sociologist. His concern is with why people say the dumb things they do, as much as with what's dumb about them, and he shows how the answer—interestingly, given the title of his books—is not only to do with duplicity. *Damned Lies and Statistics*, and now *More Damned Lies and Statistics* (both University of California Press, 2001 and 2004), have hatfuls of contemporary examples of rogue numbers, including one now famously described as the worst social statistic ever. These books are clear, thoughtful in their analysis of the way the number industry works (particularly among advocacy groups), and entertaining, and offer a good guide to more critical thinking.

Innumeracy by John Allen Paulos (Penguin, 2000) is equally replete with examples, often funny, occasionally complaining, but sometimes brilliantly imaginative, about all manner of numerical garbage. It tends toward the psychology of these errors, asking why people are so susceptible to them, and offers forthright answers. The innumeracy he has in his sights is partly an attitude of mind, which he strives to talk readers out of. The book also gives a useful reminder of how simple classroom math can be used to represent the everyday world.

Gerd Gigerenzer annoys one or two proper statisticians by preferring to be broadly intelligible than always technically correct. *Reckoning with Risk* (Penguin, 2003) does two valuable things well: it weans the reader off an attachment to certainty, and shows how to talk about risk in a way that makes more intuitive sense, even if it does cut one or two corners. We use the same method here.

Dicing with Death by Stephen Senn (Cambridge University Press, 2003) is full of wry humor and also deals with risk and chance, particularly in health. It is more technically difficult in places, and adds chunks of historical color, but is well worth the effort for those who want to begin to develop an academic interest.

Risk (UCL Press, 1995) by John Adams is mostly accessible, skillfully provocative on subjects the reader is surprised to find are not straightforward, and makes a sustained case about the nature of behavior around risk. It also includes some social theory, which won't be to everyone's taste, but the power is in the numbers.

Simon Briscoe's *Britain in Numbers* (Politico's, 2005) is an extremely useful survey of the strengths and weaknesses of a wide range of UK economic and social indicators. He has a sharp eye for a flaw and a sardonic line in political comment.

The Tyranny of Numbers by David Boyle (Flamingo, 2001) is, as the title suggests, a polemic against the fashion for measuring everything. It overstates its case, makes hay out of the woes of certain historical figures, and altogether has a great time railing at the world's reductive excesses, just as a polemic should. For fun and provocation rather than measured argument.

Two more specialized but excellent analyses are David Hand's *Information Generation: How Data Rule Our World* (Oneworld, 2007) and Michael Power's *The Audit Society* (Oxford University Press, 1997).

For a more formal, elementary introduction to statistics, or more importantly to statistical ways of thinking, *Statistics Without Tears* by Derek Rowntree (Penguin, 1988) is a good place to start, particularly for nonmathematicians.

It's worth making one exception to our rule about books on numbers in the news, with three that are worth reading as entertaining introductions to more numerical habits of mind, all by Rob Eastaway. *Why Do Buses Come in Threes?*, *How Long Is a Piece of String?*, and *How to Take a Penalty* (all from Robson Books, 2005, 2007, and 2003, respectively).

There are a number of good Web-based commentators. Chance News regularly steps into the fray on statistically related news stories: http://chance.dartmouth.edu/chancewiki/index.php/Main_Page. STATS (www.stats.org) attempts something similar, though feels less academically detached and seems to regard its main objective as challenging food and environmental health scares. John Allen Paulos has a column that is always entertaining on the *ABC News* Web site: http://abcnews.go.com/Technology/WhosCounting. There are a great many other blogs and the like that regularly find their way onto statistical stories, and of varying political persuasions, but apart from http://blogs.wsj.com/numbersguy at the *Wall Street Journal*, which is consistently levelheaded, and Marcus Zillman's useful collection of statistics resources—http://statisticsresources .blogspot.com—readers probably are best advised to find their own favorites.

INDEX

Note: Page numbers in *italics* refer to tables.